PUZZLES FOR THE HORMONES

CROSSWORDS FOR TEENAGERS

50 MEDIUM CROSSWORD PUZZLES

PUZZLE THERAPIST

CROSSWORD | SUDOKU | KIDS & ADULTS

CONTENT

PUZZLE 1

ACROSS

1. The English translation for the french word: hËme

5. In ___ (where found)

9. Fill in the blank with this word: "Agnus ___"

12. Single-named supermodel

13. Russia's Lake ___

15. Unscramble this word: dere

16. Rock guitarist once married to Goldie Hawn

18. Your highness?: Abbr.

19. Verdi's "___ tu"

20. Put back

21. They may take a few yrs. to mature

23. Lego competitor

24. The English translation for the french word: razzia

25. Salty gulp

28. Play characters

32. Sheik ___ Abdel Rahman

33. Soliloquy starter

34. Underwriter's concern

35. Vincent Lopez's theme song

36. Specks

37. TV show for which Bill Cosby won three Emmys

38. Put ___ words

39. Sorority letters

40. Fill in the blank with this word: ""Scrubs" co-star ___ Braff"

41. a book of recipes and cooking directions

43. Wynn of "Dr. Strangelove"

45. Fill in the blank with this word: """___ kleine Nachtmusik"""

46. Where Manhattan is: Abbr.

47. This 11th letter of the Greek alphabet is used to symbolize wavelengths

50. Depts. in depts.

51. Usher's offering

54. Simple rhyme scheme

55. End of a come-on to bargain hunters

58. Prefix with -itis

59. Fill in the blank with this word: ""Maid of Athens, ___ part": Byron"

60. St. Louis's historic ___ Bridge

61. Woe ___' (popular grammar book)

62. Watchdog's sound

63. Fill in the blank with this word: ""It's fun to stay at the ___": Village People"

DOWN

1. Nectar-pouring goddess

2. The ruler of Qatar is known by this 4-letter title

3. Year in the middle of this century

4. Fill in the blank with this word: "Electric ___"

5. Supreme Court justice since 1990

6. The Dow, e.g.

7. Fill in the blank with this word: "Eye ___"

8. Actor Tognazzi

9. Shoulder muscle, informally

10. Unusual shoe spec

11. Kupcinet and Cross

14. Star in Scorpius

15. Freshens, in a way

17. First name in Egyptian politics

22. Sue Grafton's '___ for Noose'

23. Where to belt one down and belt one out

24. Singer McEntire and others

25. Fill in the blank with this word: "___ the Hedgehog (video game)"

26. See 81-Across

27. Finnish architect Alvar ___

28. Chaim who wrote 'The Chosen'

29. Passover's month

30. Watchdog org.?

31. Brainteaser

33. Tough row ___

36. When coins came into general use

42. The English translation for the french word: prier

43. Julie ___, voice of Marge Simpson

44. Start of Massachusetts' motto

46. The English translation for the french word: kiowa

47. Legal scholar Guinier

48. Vigoda and Fortas

49. Windsurfers' mecca

50. Call in the game Battleship

51. William Saroyan's "My Name Is ___"

52. Relief provider

53. Fill in the blank with this word: "___ Verde National Park"

56. Work unit

57. Whimsical

PUZZLE 2

ACROSS

1. Zulu, in the military: Abbr.

4. Zest

8. The English translation for the french word: solliciter

13. Finish this popular saying: “Possession is nine points of the_____.”

14. (2003) “Pieces of ___”

16. Top

17. Formula ___

18. The English translation for the french word: sigma

19. Mark, as a trail

20. Limiting the number of children born

23. Top dog

24. Unscramble this word: atx

25. Medicine’s ___ system

28. Use a surgical beam

29. Trig. function

32. Fill in the blank with this word: “”Pumping ___””””

33. Superman player

35. Chicago suburb

37. English speaker in Africa

40. Knocking noise

41. The English translation for the french word: Lyon

42. Uh-Oh! ___ (Nabisco product)

43. This Scandinavian’s “John Gabriel Borkman” takes place on a winter night in a snowstorm

45. Finish this popular saying: “While there’s life there’s_____.”

49. The English translation for the french word: touffu

50. X-File subj.

51. The English translation for the french word: tÈmoignage

52. An unobserved and uninteracting observer

56. With lance in hand

59. Domesticated ox

60. Taro dish

61. This word refers to extreme opposites or a type of bear

62. The English translation for the french word: vanter

63. Fill in the blank with this word: “”___ true!””

64. The English translation for the french word: polka

65. The English translation for the french word: plateau

66. Was on the bottom?

DOWN

1. Unscramble this word: lglabo

2. The paper for this type of envelope was originally made from hemp of a certain Philippine city

3. Pipsqueaks

4. Turkish title for officials

5. Vast

6. Fill in the blank with this word: “”Cogito, ___ sum””

7. Sketch

8. Part of many addresses

9. The English translation for the french word: se calmer

10. Fill in the blank with this word: “___ Aquarids (May meteor shower)”

11. This axlike tool is a 3- or a 4-letter word, depending on whether you spell it with that final “E”

12. John ___

15. Fill in the blank with this word: “CaffË ___”

21. Legalese adverb

22. The Maine coon cat is named for its resemblance to this ring-tailed critter

25. Gray ___

26. whelped (similar term)

27. The recording artist now billed simply as this said, “People have so much emotion... toward the word Yoko”

29. The English translation for the french word: tÈmoigner

30. The plural of the word ovum

31. Unloads

32. Event with pairs and eights

34. Victorian ___

36. Holiday ___

37. Wrapped garment

38. Fill in the blank with this word: “Big-ticket ___”

39. Alert watchfulness: eagle ____

40. Towel holder

44. The English translation for the french word: schnick

46. Striped African animals

47. Jai alai ball

48. The English translation for the french word: recruter

50. Ne plus ___

51. Snappish

52. Fill in the blank with this word: “___ jacket”

53. Waiting room call

54. Way up?

55. Fill in the blank with this word: “___ Bowl”

56. The English translation for the french word: appli

57. Fill in the blank with this word: “”___ bad!””

58. Taken ___

PUZZLE 3

1	2	3	■	4	5	6	■	■	7	8	9	10	11	12
13			■	14			15	■	16					
17			■	18				■	19					
20			21		■	■	22	23						
24				■	25	26					■	■	■	■
■	■	27		28					■	29	30	31	32	33
34	35		■	36				■	■	37				
38			39		■	40		41	■	42				
43					■	■	44		45		■	46		
47					■	48					49		■	■
■	■	■	■	50	51					■	52		53	54
55	56	57	58					■	■	59				
60						■	61	62	63		■	64		
65						■	66				■	67		
68						■	■	69			■	70		

ACROSS

1. What the Cubs play games in: Abbr.

4. Payroll service giant, initially

7. Some steaks

13. Uncle ___ of "Seinfeld"

14. Hurdle for some univ. seniors

16. Tony winner O'Shea

17. Worker in the TV biz

18. Herder

19. What a knockout!'

20. Job extras

22. Brute strength

24. Zest

25. Painter Modigliani

27. Having digits

29. 1998 World Series star Ricky

34. Fill in the blank with this word: ""Wanna ___?"""

36. Fill in the blank with this word: "___ prof."

37. Ones with iron hands

38. Hoist ___ (enjoy the pub)

40. Fill in the blank with this word: "D.C.'s Kennedy ___"

42. Siouan speakers

43. The ___ Adventure, at SeaWorld

44. Year Trajan was born

46. Shelter grp.

47. More vigorous

48. Helpful group on-line

50. Fireplace receptacle

52. Fill in the blank with this word: "___ Mary's (L.A. college)"

55. Italian appetizer, literally "little toasts"

59. Under-the-sink fitting

60. This familiar nickname for folks from down under is a shortened & altered form of "Australian"

61. Steps down to a river, in India

64. Rock's Brian ___

65. Wound

66. In ___ (where found)

67. Word on a dipstick

68. Richard of TV's "The Real McCoys"

69. Slant

70. Give __ break!'

DOWN

1. "NYPD Blue" Emmy winner Gordon

2. Pfeiffer of TV's 'Cybill'

3. Coming in a rush

4. Some movie theaters

5. Mexican-style fast-food chain

6. What a dog might "shake" with

7. Writer of 31-Across

8. King of Belgium, 1934-51

9. Fill in the blank with this word: "___ Fjord"

10. Fill in the blank with this word: """___ Stars," #1 hit for Freddy Martin, 1934"

11. Fill in the blank with this word: "Christie's "Death on the ___"""

12. The English translation for the french word: saisir

15. Ones with a wolf at the door

21. Nickname of 1954 home run leader Ted

23. Military asst.

25. Robert Louis Stevenson's "___ Triplex"

26. Richie's mother, to the Fonz

28. coming to full development

30. Want ad inits.

31. Toward the mouth

32. Greenland base for many polar expeditions

33. She, in Italy

34. Wingding

35. Biblical dry measure: Var.

39. Fill in the blank with this word: """Trust ___" (1937 hit)"

41. Fill in the blank with this word: "Costa ___"

45. Him, in Heidelberg

48. Ending with tele-

49. Worker with a saving plan, for short

51. Tuscan city

53. Jockey great Earl ___

54. Take on

55. Math class, for short

56. Sorry soul

57. Bone: Prefix

58. Takes root

59. Spittoon sound

62. Old what's-___-name

63. Sports org. that publishes DEUCE magazine

PUZZLE 4

1	2	3	■	4	5	6	7	8	■	■	■	9	10	11
12			■	13					■	14	15			
16			■	17					18					
19			20				■	21				■	■	■
22					■	■	23				■	24	25	26
■	■	27			28	29					30			
31	32		■	33				■	■	■	34			
35			36	■	37			38	39	■	40			
41				■	■	■	42			43	■	44		
45				46	47	48					49		■	■
50			■	51				■	■	52			53	54
■	■	■	55				■	56	57					
58	59	60					61				■	62		
63					■	64					■	65		
66			■	■	■	67					■	68		

ACROSS

1. Mai ___

4. White fringe tree, named for its fringes of white flowers, is also called "old man's" this facial feature

9. Wharton grad

12. Fill in the blank with this word: “”Humanum ___ errare”””

13. Unscramble this word: laret

14. Morley of “60 Minutes”

16. Theater sound

17. Review, part 3

19. Philosopher who coined the phrase “the best of all possible worlds”

21. Whip mark

22. Fill in the blank with this word: “”Bonne ___!” (French cry on January 1)”

23. Throw your weight around practicing this martial art whose name means “gentle way”

24. Michigan's ___ Canals

27. Makes cereal more flavorful?

31. Verdi's “___ giardin del bello”

33. Suffix in nuclear physics

34. Long time: Var.

35. Wings

37. Louisiana's ___ Cajuns

40. With 121-Across, part of an afternoon repast

41. When doubled, popular 1980s-'90s British sitcom

42. Money writer Marshall ___

44. Book before Zephaniah: Abbr.

45. Public speaker's goal

50. What's funded by FICA, for short

51. Mountain ___ (Pepsi products)

52. Mexican moolah

55. Onetime Chevy subcompact

56. It may have an arch or a lintel

58. With 82-Across, two things at an amusement park

62. Stamps, say

63. Fill in the blank with this word: ""Eating ___" (1982 black comedy)"

64. UnitedHealth rival

65. Rapper ___-A-Che

66. Special ___

67. Fill in the blank with this word: "___ no"

68. Racecar driver ___ Fabi

DOWN

1. Unit used in electromagnetism

2. woody (similar term)

3. Turndown #2

4. Be on deck

5. Wells's oppressed race

6. From ___ (the works)

7. Whistler, at times

8. Draft

9. Painter's deg.

10. Tight end, at times

11. Fill in the blank with this word: ""If the ___ is concealed, it succeeds": Ovid"

14. Nail ___

15. Wont

18. The scarlet letter

20. Fill in the blank with this word: """___ Mir Bist Du Schoen" (1938 hit)"

23. Lions' "kingdoms"

24. Mascot #6

25. Florida's ___ National Forest

26. Plaint from Laura Petrie

28. Verb type: Abbr.

29. Shelley's "___ Skylark"

30. Spanish sight seer?

31. Moves toward

32. Raines and Fitzgerald

36. Wobbly walker

38. Lithium-___ battery

39. Unscramble this word: net

43. Like the earth or a bar magnet

46. Tylenol rival

47. Two out of twenty?

48. Like most streets

49. Whitman's "A Backward Glance ___ Travel'd Roads"

53. Jack of "The Great Dictator"

54. Food service giant based in Houston

55. Indonesia's ___ Islands

56. Some police officers: Abbr.

57. Wine: Prefix

58. To and ___

59. Literary monogram

60. Sigur ___ (Icelandic post-rock band)

61. Fill in the blank with this word: """___ Haw"""

PUZZLE 5

ACROSS

1. Twosome

5. Hate or fear follower

9. Five-times-a-day Islamic prayer

14. 1990 World Series M.V.P. Jose

15. Fill in the blank with this word: "Auvers-sur-___, last home of Vincent van Gogh"

16. Successor to the Cutlass

17. D.O.E. part: Abbr.

18. Explosives

19. Actor Richmond and singer Jackson

20. Chivalrous

22. Unlucky number for Caesar?

23. *Illness caused by eating Cheetos?

28. Site of 1990s genocide

32. Middle of a famous palindrome

33. Wild ___

34. Fill in the blank with this word: "___ self-defense"

35. Wee hour

38. "Faster!"

42. Suffixes with glycer- and phen-

43. Thwart in court

44. Fill in the blank with this word: "___ Grande, Ariz."

45. The English translation for the french word: ramen

46. Jeepers!' in Jersey

48. Modern medicine tablet feature

52. Fill in the blank with this word: "___ and Thummin (Judaic objects)"

53. Seine-___, department bordering Paris

58. Fill in the blank with this word: ""Tony n' ___ Wedding" (theater hit)"

60. You've Really ___ Hold on Me'

62. Something that takes its toll?: Abbr.

63. Split up

64. Give an ___ effort

65. New York's Jacob ___ Park

66. French peers

67. What P.O.'s handle

68. Finish this popular saying: "Every stick has two______."

DOWN

1. Historic Scott

2. Soprano ___ Huang

3. Leaving on ___ Plane'

4. Quad building

5. Lepidopterous movie monster

6. Where you may have a yen for shopping

7. Fill in the blank with this word: ""Humanum ___ errare"""

8. Low-___

9. Woeful

10. Soviet leader ___ Kosygin

11. Sainted ninth-century pope

12. Saxophonist ___ Lawrence

13. Fill in the blank with this word: ""___, With Love"""

21. Walloped, quickly

24. Yeah, that's good'

25. Outfit

26. The Isle of Man's Port ___

27. Ward (off)

28. Fill in the blank with this word: "El ___ (cheap cigar, in slang)"

29. make high-pitched, whiney noises

30. Lawyers: Abbr.

31. Wiretapping grp.

34. Soviet co-op

35. Way around London, once

36. Unscramble this word: siwe

37. Buckwheat's affirmative

39. North Carolina's Cape ___

40. Woe ___!'

41. Motel freebie

45. Forward

46. Bluish white twinklers

47. Fill in the blank with this word: "___ Zeppelin"

48. Student getting one-on-one help

49. Infuriating

50. Fill in the blank with this word: ""I don't ___ bit"""

51. Philanthropist Brooke ___

54. Sein : German :: ___ : French

55. It's in the back row, right of center

56. Fill in the blank with this word: ""___ you not"""

57. Series of legis. meetings

59. Ways around: Abbr.

60. Toothpaste type

61. Thomas Moore poem "___ in the Stilly Night"

PUZZLE 6

■	1	2	3	■	4	5	6	7	■	8	9	10	11	12
13				■	14				■	15				
16				17					■	18				
19					■	■	■	■	20					■
21			■	22	23	24	■	25			■	26		27
28			29				■	30			■	31		
■	■	■	32				33				■	34		
35	36	37		■	38					■	39			
40			■	41						42		■	■	■
43			■	44			■	45				46	47	48
49			■	50			■	51			■	52		
■	53		54			■	■	■	■	55	56			
57					■	58	59	60	61					
62					■	63				■	64			
65					■	66				■	67			■

ACROSS

1. Mike Ovitz's former co.

4. Old Soviet secret police org.

8. Meaning small, round & glittering, this adjective is usually applied to the eyes of the untrustworthy

13. William of TV's "The Greatest American Hero"

14. Fill in the blank with this word: "Acid ___"

15. Bony cavities, anatomically

16. bred of closely related parents

18. Spoils, with "on"

19. Fill in the blank with this word: "___ Lama"

20. See 38-Down

21. Peace Nobelist Kim ___ Jung

22. What Austrians speak: Abbr.

25. Ready to relieve 'em of a ___ or two' ('Les Mis√©rables' lyric)

26. Ozone hazard, for short

28. Each man's entitlement, per the saying

30. Fill in the blank with this word: "At ___ rate"

31. Lew Wallace's "Ben-___"

32. the 4th brightest star and the brightest star in the constellation Bootes

34. Yacht's dir.

35. Tiny bit of a tick: Abbr.

38. Tea variety

39. Warehouse contents:

Abbr.

40. Fill in the blank with this word: “1971 sci-fi film “___ 1138””

41. What ethylene may be used for

43. Some like it _______ plate

44. WSW’s reverse

45. Notre Dame cry

49. GM: “___ the USA in your Chevrolet”

50. They line up between centers and tackles: Abbr.

51. 1983 Indy winner Tom

52. Operations ___ (Army position)

53. Les __ Mousquetaires’

55. Fill in the blank with this word: “Buffalo Bill ___ Wild West Show”

57. Fill in the blank with this word: “”Smoking ___?””

58. Religious tract?

62. Lots and lots

63. Fill in the blank with this word: “Bust ___ (laugh hard)”

64. Syllables before “di” or “da” in a Beatles song

65. U.S. Grant opponent

66. Makes it

67. Fill in the blank with this word: “___ Aquarids (May meteor shower)”

DOWN

1. Where Caleb was sent as a spy

2. Potsdam Conference attendee

3. Fill in the blank with this word: “___ lot (gorged oneself)”

4. Venus or Mars

5. The English translation for the french word: lÈpisostÈe

6. Sugar ___

7. German link

8. Villains

9. Fill in the blank with this word: “”... there are evils ___ to darken all his goodness”: Shak.”

10. Like many files nowadays

11. “Moon Over Parador” star, 1988

12. Title syllables in a 1961 Lee Dorsey hit

13. The English translation for the french word: mioche

17. The English translation for the french word: rigueur

20. Lacking bargaining power, maybe

23. Artistic impressions?

24. Fixes, as a manuscript

25. Wrapped garments

27. Tribe in Manitoba

29. Mini-___

33. What’s the ___?’

35. That you should feed a cold and starve a fever, and others

36. Shape preserver, of a sort

37. On the surface

39. Team V.I.P.: Abbr.

41. Get the last bit of suds out

42. Great-___

46. Under water

47. Strait of Messina menace

48. Popular landscaping plant

54. Spy Mata ___

56. Poulenc’s “Sonata for ___ and Piano”

57. Scottish Peace Nobelist John Boyd ___

58. When Virginia got a state one of these animals, the gov.’s office released a poem that mentioned Adam West

59. Saccharin discoverer ___ Remsen

60. Worth mentioning

61. Weight abbr.

PUZZLE 7

1	2	3	4	■	5	6	7	8	■	9	10	11	12	13
14				■	15				■	16				
17				■	18				■	19				
20				21					22			■	■	■
■	■	■	23				■	24			■	25	26	27
■	28	29				■	■	■	30		31			
32			■	33		34	35	36						
37			38	■	■	39			■	■	40			
41				42	43				44	45	■	46		
47						■	■	■	48		49			■
50			■	51		52	■	53				■	■	■
■	■	■	54				55					56	57	58
59	60	61			■	62				■	63			
64					■	65				■	66			
67					■	68				■	69			

ACROSS

1. Fill in the blank with this word: "2000 Olympic hurdles gold medalist ___ Shishigina"

5. West Virginia resource

9. Gave up

14. Fill in the blank with this word: "___ oak"

15. Fill in the blank with this word: "___ in a blue moon"

16. Fill in the blank with this word: "___ fixes"

17. Winter coating

18. Michelle of "Crouching Tiger, Hidden Dragon"

19. Actress Beulah

20. Hoosier pro

23. Formal hat, informally

24. Soldier

25. Workout unit

28. They test reasoning skills: Abbr.

30. Ordered group of numbers in math

32. Permitted

33. Start of instructions for solving this puzzle

37. Mussorgsky's 'Pictures ___ Exhibition'

39. Power ___

40. Heads ___, tails...'

41. Fill in the blank with this word: ""Merci beaucoup" : France :: ___ : Japan"

46. Fill in the blank with this word: "Conductor ___-Pekka Salonen"

47. Houston pro soccer team

48. Welcome

50. Wilt

51. Alley org.

53. The English translation for the french word: ouvrage

54. 50's-90's jazz singer

59. Successor to Clement VIII

62. Unstable leptons

63. Sound

64. Salami choice

65. Penny-___ (trivial)

66. Helgenberger of "CSI"

67. Proust title character

68. Old Chinese money

69. Some TV spots, briefly

DOWN

1. It's you! What a surprise!'

2. Fill in the blank with this word: "Crazy as a ___"

3. Unscramble this word: ladg

4. Hindu drink of the gods

5. the affectation of being demure in a provocative way

6. Two semesters

7. Reynolds film "Rent-___"

8. Composer Franz

9. Symbol on an old quarter

10. Whiff

11. Fill in the blank with this word: "China's Sun Yat-___"

12. Turner of TV channels

13. Fu-___ (legendary Chinese sage)

21. Rental units: Abbr.

22. Water-to-wine site

25. Made like a geyser

26. usually good-naturedly mischievous

27. Fill in the blank with this word: ""Lovergirl" singer ___ Marie"

28. One of the Jacksons

29. With force and much noise

31. Switzerland's Bay of ___

32. Hollywood's Alan and Diane

34. Maritime CIA

35. Scott Joplin's "Maple Leaf ___"

36. Worker with books, for short

38. Fill in the blank with this word: ""...___ thousand times...""

42. Semiterrestrial organism

43. The ___ Report (upscale magazine)

44. 38-Across's real name, in brief

45. Mechanic's ___

49. How some kids spend the summer

52. Old tombstone abbr. meaning "at the age of"

53. Instrument with fingerholes

54. White matter component

55. Indian of the Sacramento River valley

56. Harem rooms

57. Peseta : Spain :: ___ : Italy

58. Yuletide quaffs

59. Some TVs and smartphones

60. Repugnant exclamation

61. Three ___ match

PUZZLE 8

1	2	3	4	5	■	■	6	7	8	9	■	10	11	12
13					■	14					■	15		
16					17						■	18		
■	■	■	19				■	■	20		21			
■	22	23		■	24		25	■	26				■	■
27			■	28				29		■	30		31	■
32			33			■	34			35				36
37					■	38			■	39				
40					41			■	42					
■	43			■	44			45			■	46		
■	■	47		48		■	49			■	50			■
51	52					■	■	53		54		■	■	■
55			■	56		57	58					59	60	61
62			■	63					■	64				
65			■	66				■	■	67				

ACROSS

1. Witch of ___

6. Curtainlike partitions, biologically

10. The Missing Drink : High ____ rose

13. Orange ___

14. Young pigeon

15. School days initials

16. Economic woe

18. Worker in the TV biz

19. To laugh, to Lafayette

20. Fill in the blank with this word: "19th-century Swedish writer Esaias ___"

22. Oh, ___ cryin' out loud!'

24. Sir ___ McKellen (Gandalf portrayer)

26. Some flawed mdse.

27. Magic, on scoreboards

28. "Bewitched" witch

30. Kwik-E-Mart clerk on "The Simpsons"

32. Time's 1981 Man of the Year

34. Way to one's heart?

37. Fill in the blank with this word: "Everything ___ place"

38. Robertson of CNN

39. What many incumbents do

40. The English translation for the french word: extatique

42. Ford from long ago

43. Source of some rings

44. "Riverdance" composer Bill

46. Singer ___ King Cole

47. Young lady of Sp.

49. Ugly as ___

50. Ways around: Abbr.

51. Remove plumbic traces from

53. Fill in the blank with this word: "___ nut"

55. Fill in the blank with this word: "___-Locka, Fla."

56. Orioles hurler (1966 champs) / Solo crooner of "Oh! My Pa-Pa" (#1 in 1954)

62. Ending with tele-

63. The second part missing in the author's name ___ Vargas ___

64. Fill in the blank with this word: "___ fatale"

65. Woeful

66. Fill in the blank with this word: "___ Arnold's Balsam (old patent medicine)"

67. Writing by Montaigne

DOWN

1. Globe: Abbr.

2. TV actress Susan

3. British verb ending

4. Fill in the blank with this word: ""And their lies caused them ___": Amos 2:4"

5. One of the Spice Girls

6. Owns

7. Pins and needles' place

8. Vientiane native

9. Fill in the blank with this word: ""Li'l ___" (Al Capp strip)"

10. What 3M's Scotch is a brand of

11. Mahler's "Das Lied von der ___"

12. Pirate's interjection

14. Three-time Masters winner

17. Fill in the blank with this word: "___ Spalko, Indiana Jones villainess"

21. Lent beauty to

22. Bygone Spanish dictator

23. National monument site since 1965

25. They're learning the ropes

27. Tot's injury

28. She, in Italy

29. Fill in the blank with this word: "___ room"

31. Palate parts

33. Italian auto maker Bugatti

35. Fill in the blank with this word: "Elvis ___ Presley"

36. Sci. course

38. Wellness org.

41. Silly talk

42. Fill in the blank with this word: "___ the world"

45. Fill in the blank with this word: ""___ Rock" (Bob Seger hit)"

48. Asian weight units

50. MS. enclosures

51. Specks

52. Biblical dry measure: Var.

54. Fill in the blank with this word: ""That's ___""

57. On the ___

58. Suffix with bull or bear

59. Horatio Nelson's ___ Victory

60. Fill in the blank with this word: "___ Savahl (couture label)"

61. Retailer with stylized mountaintops in its logo

PUZZLE 9

1	2	3	4	■	■	5	6	7	8	■	9	10	11	12
13				■	14					■	15			
16				17						■	18			
19			■	20				■	21	22				
23			24			■	■	25				■	■	■
■	■	■	26			27	28					29	30	31
32	33	34		■	■	35					■	36		
37				■	■	38			■	■	39			
40			■	41	42				■	■	43			
44			45						46	47		■	■	■
■	■	■	48				■	■	49			50	51	52
53	54	55				■	56	57			■	58		
59				■	60	61					62			
63				■	64					■	65			
66				■	67				■	■	68			

ACROSS

1. Fill in the blank with this word: "Club ___ (resorts)"

5. Sound of an exploding cigar

9. Shadow

13. Trapped like ___

14. Stomach soother, for short

15. Vintage vehicles

16. Wedding party

18. Greek gulf or city

19. Wall Street earnings abbr.

20. Officially listed: Abbr.

21. T. Boone Pickens, for one

23. Like the laws of kosher food

25. Year Queen Victoria died

26. Prestige of Jay's predecessor?

32. Livin' La Vida ___' (Ricky Martin hit)

35. Using gold jewelry from the Israelites, he fashions the Golden Calf

36. Fill in the blank with this word: “”Should ___ shouldn't ...””

37. Takeoff artist

38. By way of: Abbr.

39. Fruit holder

40. Wagner's "___ fliegende Hollander"

41. "Yeah"

43. Beginning

44. Something delicious to drink

48. Fill in the blank with this word: "___ Phillips, who played Livia in "I, Claudius""

49. What a knockout!'

53. N.Y.C. thoroughfare in the Rodgers and Hart song "Manhattan"

56. Asian, e.g.

58. Fill in the blank with this word: "Elevator ___"

59. Seuss's "Horton Hears ___"

60. Emotional scene with actor Grant?

63. Wine vintage

64. Wide collars

65. Zoroastrian spirit

66. Fill in the blank with this word: """Zip-___-Doo-Dah""

67. Well, fiddle-dee-dee! He was the fifth emperor of Rome

68. Zip

DOWN

1. Pot

2. Verdi baritone aria

3. Role-playing game, briefly

4. Brit. money

5. Toot one's own horn

6. Fill in the blank with this word: """___ of the Flies""

7. French soul

8. Home of the Atlas Mountains

9. Language family that includes Finnish and Hungarian

10. The English translation for the french word: berme

11. Italian composer Nino ___

12. Wise ___ owl

14. General ___ chicken

17. Vivacity

22. Fill in the blank with this word: """___ Ordinary Man" ("My Fair Lady" song)"

24. Fill in the blank with this word: """When it's ___" (old riddle answer)"

25. Some aromatic resins

27. Lane with lines

28. Prophet who predicted the destruction of Nineveh

29. That ole boy's

30. Years on end

31. Fill in the blank with this word: "Family ___"

32. Fill in the blank with this word: "___-da (pretentious)"

33. Well-running group?: Abbr.

34. Smear with wax, old-style

39. West Virginia resource

41. "Trinity" novelist

42. Missionary's target

45. Where some errands are run

46. Chaplin and others

47. Fill in the blank with this word: """The moon is ___; I have not heard the clock": "Macbeth""

50. Gradually quickening, in mus.

51. The English translation for the french word: larve

52. The English translation for the french word: aryen

53. Poet Angelou

54. Was in the red

55. Those, to Robert Burns

56. Haing S. ___ (Oscar winner for 'The Killing Fields')

57. Speed skater Apolo Anton ___

61. Sport-___ (vehicle)

62. Potus #34

PUZZLE 10

ACROSS

1. Mid fourth-century year

6. The English translation for the french word: lasso

11. Vous √™tes ___'

14. Indonesian island

15. Wicked Game' singer Chris

16. Bhikkhuni : Buddhism :: ___ : Catholicism

17. Calisthenics for show-offs

19. Some accounting entries: Abbr.

20. Public Theater founder

21. Speedometer reading: Abbr.

22. Suffix with social

23. TMZ fodder

26. This "Y" word is in the title of Robert Crawford's 1939 Air Corps (now Air Force) song

28. Supermodel Carol

29. No longer hindered by

33. Kipling's 'Follow Me ___'

34. Worldwide workers' grp.

35. Okinawa port

36. Tithing portion

39. The luxo, with springs to tilt it at the correct angle, is a favorite desktop one of these

41. Pa's pa

43. Remarkably, in commercialese

44. Fill in the blank with this word: "Belgian violin virtuoso Eugene ___"

46. Fill in the blank with this word: "Family ___"

47. Where Schwarzenegger was born: Abbr.

48. TV screen: Abbr.

49. Painter Andrea del ___

51. W.W. II vessel: Abbr.

52. Partnership

55. Totals

57. Harem room

58. When H

60. Sub ___ (secretly)

61. S.P.C.A. part: Abbr.

62. Hint to 17-, 28- and 43-Across

67. Ticket abbr.

68. Sportscaster Rashad

69. Tropical vine

70. Taxonomic suffix

71. The English translation for the french word: attente

72. U.S.M.C. noncoms

DOWN

1. Year in the life of Constantine

2. O-___-O (brand of sponge)

3. Neckcloth

4. They're pulled in pulkas

5. The English translation for the french word: dÈtÈriorer

6. The English translation for the french word: zÈzayer

7. Fill in the blank with this word: "Carolina ___"

8. Fill in the blank with this word: "___ Centre, Minn. (Sinclair Lewis's birthplace)"

9. Mawkish

10. Response to "I have a question for you"

11. Phone calls, room service charges, etc.

12. Say "@#$%!"

13. Wordless song: Abbr.

18. Ante action

23. With glee

24. Stewpots

25. an ache localized in the stomach or abdominal region

27. White Sands Natl. Monument state

30. The 4 sections attached to these to help them go straight in barrooms are called flights

31. Scarlett ___ of "Gone With the Wind"

32. Hall-of-___

37. Fill in the blank with this word: ""It is equally an error to ___ all men or no man": Seneca"

38. Word repeated before some relatives' names

40. Nickname for JosÈ

42. Fill in the blank with this word: "Alexander ___, Russian who popularized a chess opening"

45. Shelley was one

50. Diving birds: Var.

52. Fill in the blank with this word: "___ Girl"

53. Vote to accept

54. The English translation for the french word: dÓme

56. Nostril

59. Salinger's 'For ___ - With Love and Squalor'

60. Some wines

63. X-ray spec?

64. Wilt

65. Treebeard, e.g.

66. Fill in the blank with this word: ""___ Kapital""

PUZZLE 11

ACROSS

1. Circus reactions

5. Pats on the back?

10. Medieval romance tale

14. Would ___?'

15. Unscramble this word: torpo

16. Fill in the blank with this word: "___ Aarnio, innovative furniture designer"

17. *Creation made with a bucket and shovel

19. Writer's supply: Abbr.

20. Where to spend birr

21. The English translation for the french word: AÁores

23. U.K. honour

24. Variety listing

25. Thought: Prefix

27. Role played by Drew Barrymore in a 1993 TV movie based on real life

32. Moroccan toppers: Var.

33. This card deck's Minor Arcana has 14 cards in each suit; a page is between the 10 & jack

34. WSW's reverse

35. TV co-star of Richard Belzer

36. Biblical epic

37. “Man oh man!”

38. ‘My mama done ___ me ...’

39. Overseas diplomat in N.Y.C., say

40. Thin as ___

41. Drinking and dancing instead of sleeping?

43. Fill in the blank with this word: “___-mell”

44. Written reminder

45. Yellow ___

46. The Preserver, in Hinduism

49. Windward

54. Time-share unit

55. Name on a plastic container

57. Sharon of ‘Dreamgirls’

58. Fill in the blank with this word: “”He fain would write ___”: Browning”

59. Writer ___ St. Vincent Millay

60. Where Is the Life That Late ___?’ (‘Kiss Me, Kate’ song)

61. Summa cum ___

62. Scottish rejections

DOWN

1. Unscramble this word: siwe

2. Wooden piece

3. Chi follower

4. Lees

5. Recording session need

6. Bears: Lat.

7. Italian composer Nino ___

8. Neighbor of Ger.

9. Represent, in a way

10. Fixes, as some fairways

11. Unscramble this word: ware

12. Fill in the blank with this word: “Family ___”

13. Fill in the blank with this word: “___ stick (incense)”

18. The bug partly stems from 6-digit dates in this 1960s “business-oriented” computer language

22. Time: Ger.

24. Mawkish

25. Difference between the rich and the poor

26. Spoke at length about, with ‘on’

27. Pied-___

28. “Olympia” painter

29. The Science of Logic’ author

30. TV actress Georgia

31. Officially listed: Abbr.

32. In ___ (where found)

36. Note offering good advice for life?

37. Heroes

39. Stratford-___-Avon

40. Fill in the blank with this word: “”...partridge in ___ tree””

42. Waiting, in a way

45. The Beatles’ “You Won’t ___”

46. Frankie with a falsetto

47. Where the Althing sits: Abbr.

48. Union member

49. Leigh Hunt’s “___ Ben Adhem”

50. It may rock you to sleep

51. Once I ___ secret love...’

52. Fill in the blank with this word: “”___ kleine Nachtmusik””

53. Vitamin bottle info

56. Work ___ sweat

PUZZLE 12

1	2	3	4		5	6	7	8		9	10	11	12	13
14					15					16				
17				18					19					
20						21						22		
23				24	25			26			27			
		28	29				30					31	32	33
			34							35				
36	37	38			39			40	41		42			
43				44			45			46				
47					48	49						50		
			51					52				53	54	55
56	57	58		59			60			61	62			
63			64						65					
66						67					68			
69						70					71			

ACROSS

1. Fill in the blank with this word: “”Look at me, ___...”””

5. Winston Churchill’s “___ Country”

9. Photo finish?

14. Unlikely to bite

15. What knows the drill, for short?

16. Waste

17. Something shown off on a half-pipe

20. Fill in the blank with this word: “___ home (out)”

21. Set straight

22. Union activist Norma ___ Webster

23. Fill in the blank with this word: “___ pardo (grizzly, in Granada)”

24. WWII intelligence org

26. Fill in the blank with this word: “”Dial ___ Murder”””

28. Area

34. Prefix with sclerosis

35. *Group with the 2000 #1 hit “It’s Gonna Be Me”

36. Three-quarters of M

39. Bride, in Brescia

42. Old laborer

43. Their feathers are used in Maori costumes

45. Try to strike

47. COLONIAL CHARM!

51. Fill in the blank with this word: "___ Galerie (Manhattan art museum)"

52. Tin ___

53. Pro-___ (some tourneys)

56. Dog doc

59. When clocks are changed back from D.S.T. in the fall

61. Light ___

63. Angry rabbits in August?

66. Symbols on old manuscripts

67. Michelle of "Crouching Tiger, Hidden Dragon"

68. Richie's mother, to the Fonz

69. Like some tattooed characters

70. Votes against

71. Vis-a-vis

DOWN

1. Fill in the blank with this word: """___ skin off my nose!"""

2. Speedy sharks

3. Stuff in a bomb

4. Fill in the blank with this word: "___ time limit"

5. Wall St. worker

6. Suspenseful sound

7. essential oil or perfume obtained from flowers

8. Fritz the Cat's creator

9. Was natural and unrestrained, slangily

10. To ___ is human ...'

11. Pinot ___ (wine)

12. The ___ Trail (route through Peru)

13. The English translation for the french word: cokÈfier

18. Wide collars

19. Thumb one's nose at

25. Three-stripers: Abbr.

27. Meet, as expectations

29. Windblown

30. To-do

31. Prefix with function

32. W.W.F. airer

33. Therefore

36. Longtime Chicago Bears coach

37. The Reds, on scoreboards

38. 100 lbs.

40. Order to Rex

41. The U.N.'s Kofi ___ Annan

44. of or relating to or composed of fat

46. Log holder

48. Anna Leonowens, e.g., in "The King and I"

49. Certs ingredient

50. Floors

54. Famous blonde bombshell

55. Food service giant based in Houston

56. Restaurateur Toots

57. Fill in the blank with this word: "___-Jutsu (Japanese martial art)"

58. Call in the game Battleship

60. Ralph Vaughan Williams's "___ Symphony"

62. Low-cost home loan corp.

64. Year in the reign of Antoninus Pius

65. Sounds of hesitation

PUZZLE 13

1	2	3	4	5	■	6	7	8	9	■	10	11	12	13
14					■	15				■	16			
17					■	18				■	19			
20					21					■	22			
■	■	■	■	23			■	24		25				
26	27	28	29				30	■	31			■	■	■
32				■	33			34	■	35		36	37	38
39				■	40				41	■	42			
43				44	■	45				■	46			
■	■	■	47		48	■	49			50				
51	52	53				54	■	55			■	■	■	■
56				■	57		58				59	60	61	62
63				■	64				■	65				
66				■	67				■	68				
69				■	70				■	71				

ACROSS

1. "Battlestar Galactica" commander

6. Wildcat with tufted ears

10. Okinawa port

14. Fill in the blank with this word: ""___ your life!""

15. Valentine for Val

16. U.K. awards

17. What a soldier wears that has a serial no.

18. Fought

19. Tot's tote

20. Solitaire game of matching pairs of cards

22. Film producer ___ Al-Fayed

23. Celtic sea god

24. South America's smallest country

26. Winter warmer

31. Winery vessel

32. Russia's ___ Mountains

33. Campus 100 miles NW of L.A.

35. They pop up in the morning

39. Without ___ (pro bono)

40. Nolan Ryan, notably

42. The last Mrs. Chaplin

43. Indonesian island

45. Young lady of Sp.

46. One of six allowed to an N.B.A. player

47. Reactor overseer: Abbr.

49. Crayfish dish

51. European annual with pale rose-colored flowers

55. Overseas bar deg.

56. Lhasa ___ (dog)

57. Delusions

63. You might go for a spin in it

64. Like flicks seen without special glasses

65. Mastodon trap

66. They have plans, for short

67. White house: Var.

68. New ___ (Congolese money)

69. Book part

70. Hoofbeat

71. Summer camp shelter

DOWN

1. Spiritedly, in music: Abbr.

2. Fill in the blank with this word: "___ bird"

3. Memo abbr.

4. Water under the bridge

5. Texas' ___ State University

6. Volcanic spew

7. the earth from his flesh

8. Nobelist Bohr

9. Deleted

10. Dozing

11. Fill in the blank with this word: "___ to mankind"

12. Fill in the blank with this word: "___ grudge (harbored resentment)"

13. Fill in the blank with this word: ""There's ___ chance of that""

21. The appendix extends from it

25. Poe's "The Murders in the ___ Morgue"

26. Sen. McCarthy's grp.

27. Ornamental pond fish

28. Six-foot vis-

29. Solve a problem by starting over

30. Fill in the blank with this word: ""It's ___ bet!""

34. Reserved

36. This can be a careless mistake or a foolish person who might make one

37. When expected

38. Unscramble this word: esla

41. London lockups

44. ___ Field

48. Unscramble this word: itcric

50. Reddish-brown

51. Many churchgoers: Abbr.

52. Crankcase part

53. Actor Sam

54. Mediterranean resort island

58. Fill in the blank with this word: "___ contendere"

59. Tiny battery

60. Windy City paper, with "The"

61. Jacquerie

62. Tucson-to-New Orleans route

PUZZLE 14

1	2	3	4	5	6	7	■	■	8	9	10	11	12	13
14							15	■	16					
17								18						
19				■	■	20			■	■	21			
■	■	■	■	22	23		■	24	25	26	■	27		
28	29	30	31				32				33		■	■
34						■	35						36	■
37				■	■	38			■	■	39			40
■	41			42	43			■	44	45				
■	■	46						47						
48	49		■	50			■	51			■	■	■	■
52			53	■	■	54	55		■	■	56	57	58	59
60				61	62				63	64				
65						■	66							
67						■	■	68						

ACROSS

1. Encamp

8. Unscramble this word: cmulse

14. Rubber

16. Decline an invitation

17. Elizabeth I in 1600

19. ZIP code 10001 locale: Abbr.

20. U.S.D.A. part: Abbr.

21. Marked, as a questionnaire box

22. Year the emperor Decius was born

24. Worker with books, for short

27. Fill in the blank with this word: "___ pardo (grizzly, in Granada)"

28. Question that's a classic pickup line

34. Town in County Kerry

35. Extreme Atkins diet credo

37. 40's theater director James

38. Year Marcian became emperor

39. Many a holiday visitor / Bandit

41. Kin of "Sacre bleu!"

44. Fill in the blank with this word: "___ gland (melatonin secreter)"

46. Cockpit gauge figure

48. Suffix with bull or bear

50. Smart ___ (wise guy)

51. Zogby poll partner

52. Wedding wear

54. Surveyor's dir.

56. Sportswear brand

60. Pastry chef creations ... and a hint to 12 other answers in this puzzle

65. Pixel pattern producer

66. Rossini opera

67. Three-footers

68. Reply from a polite young'un

DOWN

1. whelped (similar term)

2. Poker star Phil

3. Way to one's heart

4. Wild revelry

5. Swabs' grp.

6. Fill in the blank with this word: "___ tuna"

7. Blow-dryer brand

8. Fill in the blank with this word: "___ juice (milk)"

9. WB competitor

10. Record label whose house band was Booker T. & the M.G.'s

11. Stuffed with ham and cheese and then saut

12. Fill in the blank with this word: "Anno ___ (in the Year of Light)"

13. ___ jazz (fusion genre)

15. Hard-boiled ____ foo yong

18. Hipster's persona

22. Revolutionary Guevara

23. The English translation for the french word: cÈ

25. U.S.M.C. rank

26. Week-___-glance calendar

28. Whse. unit

29. Utah's "Family City U.S.A."

30. Painfully aware people?

31. "Nadja" actress L

32. Fill in the blank with this word: ""This ___" (shipping label)"

33. TV's Gray and Moran

36. Coal-rich German region

38. Used a thurible on

40. Corduroy ridge

42. The English translation for the french word: DBA

43. Sue Grafton's '___ for Innocent'

44. Zip

45. Woe ___' (popular grammar book)

47. Takeoff site

48. Bit of I.C.U. equipment

49. This golden-brown tint from a marine animal is applied to black & white images to give them an antique look

53. Whitewash

55. Turing test participant

56. Unscramble this word: marf

57. Unb

58. The English translation for the french word: LÈda

59. Fill in the blank with this word: ""___ sure you know...""

61. The Mavericks, on scoreboards

62. They're longer than singles, briefly

63. Wichita-to-Omaha dir.

64. What secondary recipients of e-mails get

PUZZLE 15

1	2	3	4	■	■	5	6	7	■	8	9	10	11	12
13				■	14				■	15				
16				■	17				■	18				
19				20					21					
■	■	■	22				■	■	23			■	■	■
24	25	26				■	27	28		■	29	30	31	32
33					■	34				35	■	36		
37					38						39			
40			■	41					■	42				
43			44	■	45			■	46					
■	■	■	47	48		■	■	49				■	■	■
50	51	52				53	54					55	56	57
58					■	59				■	60			
61					■	62				■	63			
64					■	65			■	■	66			

ACROSS

1. Word on a gift tag

5. Wisc. clock setting

8. Fill in the blank with this word: "Ann ___, Mich."

13. The English translation for the french word: LÈda

14. Sinatra's "Meet Me at the ___"

15. Baseball general manager Billy

16. U.S. ___ (annual Queens event)

17. Rara ___

18. Fill in the blank with this word: ""Wie geht es ___?" (German greeting)"

19. Hopper

22. Poke-___!' (kids' book series)

23. They dug his grave ___ where he lay': Sir Walter Scott

24. Potential checkout correction

27. The English translation for the french word: Hel

29. Mythical mother of the Titans

33. Michigan, e.g., to a Spaniard

34. Fungus, in Falmouth

36. Sue Grafton's "___ for Lawless"

37. End of the quip

40. Ka ___ (Hawaii's South Cape)

41. Radiohead singer Thom

42. Five, on a gunslinger's gun

43. Union Sq. and Times Sq.

45. Work units: Abbr.

46. Urges

47. TV schedule abbr.

49. Thai tender

50. Lawyers' requests at trials

58. Fill in the blank with this word: "Ad ___ per aspera (Kansas' motto)"

59. Pres. appointee

60. Zip

61. San Francisco's Museo ___ Americano

62. Trig function

63. Golf innovator Callaway and bridge maven Culbertson

64. Swindler

65. Weapons prog. since 1983

66. Zaire's Mobutu ___ Seko

DOWN

1. Whip

2. Second ed.

3. Performance halls

4. Overly zealous

5. The English translation for the french word: convoiter

6. Fill in the blank with this word: "___ and Span"

7. Fill in the blank with this word: ""Don't ___ me, bro!"""

8. Take ___!' ('Try some!')

9. Move, as a picture

10. Judge's seat

11. Fill in the blank with this word: ""The ___ lama, he's a priest": Nash"

12. Oscar-winning French film director ___ Cl

14. Star of "Gigi" and "Lili"

20. City southeast of Roma

21. Capital of ancient Macedonia

24. Rational and irrational numbers

25. Convocation of witches

26. Dutch painter Jan

27. Traffic sounds

28. World chess champ, 1935-37

30. When ___ said and done

31. Trailer for farm animals?

32. Words after ugly or guilty

34. On your ___

35. Big scrap

38. Tom Harkin, for one

39. Parts of some Mediterranean orchards

44. Fill in the blank with this word: """___ Nacht" (German Christmas carol)"

46. Snap course

48. Fill in the blank with this word: "___ chocolates"

49. Lawn game

50. Like a romantic evening, maybe

51. "@#$%!," e.g.

52. Pad ___ (noodle dish)

53. Sound of a leak

54. Ticked (off)

55. Fill in the blank with this word: ""Winnie ___ Pu"""

56. Strikeout symbols, in baseball

57. Start of Massachusetts' motto

PUZZLE 16

ACROSS

1. Natl. Boss Day, ___ 16

4. Terrif

7. World view

10. Satellite ___

13. It premiered the day before “E.R.”: “___ Hope”

15. Muhammad ___

16. Fill in the blank with this word: “Electric ___”

17. Container for preserving historical records

19. Unruly crowd

20. Unrushed pedestrian

21. Exaggerate

23. The English translation for the french word: sinople

24. The English translation for the french word: fidÈlitÈ

28. Fill in the blank with this word: “”Able was I ___ Ö””

29. The English translation for the french word: moi

30. Observation

31. The English translation for the french word: imago

33. “Listen!”

34. Things to “see” in an encyclopedia

40. Fill in the blank with this word: “___ of Sandwich”

41. Threefold

42. Untrustworthy types

45. Fill in the blank with this word: “___ and outs”

46. Unscramble this word: opp

49. Any nonverbal action or gesture that encodes a message

52. Water carrier

53. Yo, this pope III had a "Rocky" one-year reign from 884 to 885

54. This job can also mean to fashion your actions to the needs of another

56. Zine reader

58. A public relations person

60. The English translation for the french word: frangin

61. Sound of contempt

62. The English translation for the french word: roucouyer

63. Fill in the blank with this word: """___ true!"""

64. The work of Kim Philby & "Harriet", or to catch sight of suddenly

65. Worker in a garden

66. Finish this popular saying: "Like father, like_____."

DOWN

1. The English translation for the french word: octave

2. Bellman

3. This quality of sound distinguishes it from other sounds of the same pitch & volume

4. Finish this popular saying: "Monday's child is fair of_____."

5. Petri dish gel

6. Thwack

7. The English translation for the french word: violet

8. The English translation for the french word: allÈguer

9. Type genus of the Pieridae

10. One who may get dispossessed?

11. Fill in the blank with this word: "___-noir (modern film genre)"

12. Vatican vestment

14. The English translation for the french word: Celte

18. Fill in the blank with this word: "Costa del ___"

22. The English translation for the french word: daphnÈ

25. Ticket choice

26. Worry

27. The "E" of B.P.O.E.

29. Germany's Dortmund-___ Canal

31. Finish this popular saying: "No man is an_____."

32. Fill in the blank with this word: """How ___ Has the Banshee Cried" (Thomas Moore poem)"

34. British tax

35. The English translation for the french word: rani

36. Aromatic Eurasian perennial

37. Fill in the blank with this word: """___ go bragh"""

38. The English translation for the french word: sonnerie

39. U.S.N.A. grad

43. Limestone regions with deep fissures and sinkholes

44. The English translation for the french word: bÈvue

46. Flap raisers?

47. Port gets its name from this second-largest Portuguese city that's 2 letters longer

48. Unscramble this word: osenpr

50. The English translation for the french word: laÔcat

51. Holiday ___

52. Fill in the blank with this word: "___ bread"

55. Unpopular spots

56. Without a Trace' org.

57. Fill in the blank with this word: """If the ___ is concealed, it succeeds": Ovid"

59. Tiny bit

PUZZLE 17

ACROSS

1. Fill in the blank with this word: "___-bitsy"

5. Turkish title

9. Fill in the blank with this word: """The Daughters of Joshua ___" (1972 Buddy Ebsen film)"

13. Okinawa port

14. What an A is not

15. Prussian lancer

16. School board output?

18. Unyielding

19. Can't do without

20. Zealous

21. Whittier war poem "Laus ___"

22. Used a BlackBerry, perhaps

23. Notion of an underwater creature?

28. Unc's wife

29. "The Producers" role

30. There's ___ in team'

33. Whipped up

34. Rocky debris

36. To heat water to 212 degrees

37. Unit of conductance

38. Stagecoach robber Black ___

39. Von Trapp girl who's "sixteen going on seventeen"

40. Subject of some sightings

43. The ___ Band, with guitarist Little Steven

46. Suffix with Mozart

47. Wrestling moves

48. Horrid

53. Tarantula-eating animal

54. TV series that premiered in 1974

55. Fill in the blank with this word: ""I've ___!""

56. Ancient reveler's "whoopee!"

57. Fill in the blank with this word: "___ ease"

58. Smart-___

59. Funnyman Foxx

60. Rousing cheers

DOWN

1. Go up: Abbr.

2. Chevrolet model

3. Fill in the blank with this word: "___ attack"

4. Manchurian border river

5. Physicist Sakharov

6. Attach, in a way

7. Fill in the blank with this word: "Amerada ___ (petroleum giant)"

8. Fill in the blank with this word: ""If the ___ is concealed, it succeeds": Ovid"

9. Ballpark fare

10. Like some pond life

11. Pool table fabric

12. Wrapped up

15. University of Illinois locale

17. Kind of pool or ride

20. Missed by ___ (was way off)

22. George Sand's "___ et lui"

23. With 69-Across, 1930s-'50s bandleader

24. Thick-shelled seafood selection

25. Word processor command

26. Suffix with social

27. Author Vonnegut and others

30. Fill in the blank with this word: ""The first ___, the angel did say ...""

31. Fill in the blank with this word: "Auvers-sur-___, last home of Vincent van Gogh"

32. Not well

34. Reserves

35. Lit ___ (college course, slangily)

36. Twice, in music

38. Screwed up big-time

39. Tv Sidekicks : Squiggy

40. X-rated

41. Vegetable container

42. Wheelchair-accessible

43. ___ Sketch (classic drawing toy)

44. The English translation for the french word: plature

45. Unscramble this word: edart

48. Fill in the blank with this word: ""___ Barry Turns 40" (1990 best seller)"

49. Sound

50. Word that can follow the ends of 17-, 26-, 43- and 58-Across

51. Singer Lovett

52. Guesses: Abbr.

54. Old what's-___-name

PUZZLE 18

1	2	3	4	■	5	6	7	8	■	9	10	11	12	13
14				■	15				■	16				
17				■	18				■	19				
20				21					22					
■	■	■	■	23			■	■	24			■	■	■
25	26	27	28				29	30				31	32	33
34					■	■	35				■	36		
37				■	38	39				■	40			
41			■	42				■	■	43				
44			45					46	47					
■	■	■	48			■	■	49			■	■	■	■
50	51	52				53	54				55	56	57	58
59					■	60				■	61			
62					■	63				■	64			
65					■	66				■	67			

ACROSS

1. Toothpaste ingredient

5. Fill in the blank with this word: "Feel the ___"

9. Fill in the blank with this word: "Arthur Miller play "___ From the Bridge""

14. Ralph Vaughan Williams's "___ Symphony"

15. High school subj.

16. Fill in the blank with this word: "Bottom of the ___"

17. Fill in the blank with this word: "1995 Physics Nobelist Martin L. ___"

18. Campus 100 miles NW of L.A.

19. Winged

20. 1962 Mitchum/MacLaine film

23. Fill in the blank with this word: "___-turn"

24. Meditation sounds

25. Tendency not to panic

34. Mature insect

35. Wildcats' org.

36. West Coast airport inits.

37. Hop ___!'

38. Contemptible sneaks

40. Memo abbr.

41. Northwest Terr. native

42. Natal native

43. Fill in the blank with this word: "___ Coyote"

44. Thomas Jefferson was the first to lead it

48. Trucial States, today: Abbr.

49. Wreath

50. Tennyson's 12-poem series

59. Painting surface

60. Fill in the blank with this word: "Elaine ___, George W. Bush's only labor secretary"

61. Tiny battery

62. Fill in the blank with this word: ""Let ___ Cake""

63. Wrench

64. Pre-___ (take the place of)

65. To everything there is a season, including these endangered birds: the least, roseate & California least

66. Vaulted space

67. One who waits in ambush

DOWN

1. Bill Clinton, for one: Abbr.

2. Compass points (seen spelled out in 20-, 26-, 43- and 53-Across)

3. Fill in the blank with this word: "___ Aarnio, innovative furniture designer"

4. Schumacher of auto racing

5. Classic 1896 Alfred Jarry play

6. a gaping grimace

7. Fill in the blank with this word: ""Oh my ___!""

8. Wittenberg's river

9. Haemoglobin deficiency

10. Most repellent

11. Much may follow it

12. Jazz singer ___ James

13. That was exhausting!'

21. Japanese folk music with a swing feel

22. Family of Slammin' Sammy

25. Uses as a source

26. Fill in the blank with this word: ""It's ___ Unusual Day""

27. Space traveler of 1957

28. Place runners?: Abbr.

29. Starts to raise, as a hem

30. What a constant hand-washer probably has, for short

31. Organic fiber

32. Fill in the blank with this word: ""Come here ___?""

33. Trio, squared

38. Some ranchers

39. Pale ___

40. Fill in the blank with this word: ""Ready, ___...!""

42. Nuts, so to speak

43. Fancy invitation specification

45. Resident of Oklahoma's second-largest city

46. Where to find grooms

47. Whet again

50. Fill in the blank with this word: ""___ Around" (#1 Beach Boys hit)"

51. Greek township

52. Fill in the blank with this word: "Battle of the ___, opened on 10/16/1914"

53. Two times tetra-

54. Glove

55. I Lost It at the Movies' writer Pauline

56. Neil Diamond's "___ Said"

57. The English translation for the french word: nuque

58. Gainesville athlete

PUZZLE 19

ACROSS

1. The “R” of R.F.D.

6. Blood: Prefix

10. Writer Deighton

14. Heavens: Prefix

15. Writing on the wall

16. Russia’s ___ Mountains

17. Like some stickers

20. Unscramble this word: iphs

21. Early capital of Georgia

22. Fill in the blank with this word: “”There is no ___ team”””

23. They’re often pressed for cash

24. Fill in the blank with this word: “___ Haute, Ind.”

28. Tony’s boss on “Who’s the Boss?”

30. Hierarchies

32. Where NaCl is collected

35. Mil. unit

36. Plea for a TV cop?

40. Seasoning in French onion soup

41. U.S.S. Enterprise defenses

42. February 4th, to some?

45. Sharpness gauge

49. Fill in the blank with this word: “”All I Ever Need ___” (Sonny & Cher hit)”

50. Tests for srs.

52. Ooh and ___

53. Like winter, vis-

56. Fill in the blank with this word: “”For here ___ go?””

57. Monarchy or parliamentary democracy

61. Phi ___ Kappa

62. Fill in the blanks with these two words: “___ time”

63. Kitty ___ (mistress in Irish history)

64. Wing: Abbr.

65. Tu-144 and others

66. Petty officers, informally

DOWN

1. Seller of Alaska, 1867

2. Sea ___

3. Fill in the blank with this word: “___ to go”

4. Take ___ (snooze)

5. Unscramble this word: lto

6. Yawn-inducing

7. Slate and Salon

8. Fill in the blank with this word: “Dessert ___”

9. Fill in the blank with this word: “”No ifs, ___ …””

10. “Parent” of 17-Across

11. Fill in the blank with this word: “___ Onassis, Jackie Kennedy’s #2”

12. Fill in the blank with this word: “1941 film “A Yank in the ___””

13. Suffix with differ

18. What sets things in motion

19. Naut. direction

23. Zoological wings

25. Skin: Suffix

26. Bring (up) from the past

27. Winter hrs. in Bermuda

29. Upper-left key

30. Whitewash

31. Where oils are produced

33. Sch. staffer

34. Taro dish

36. Visitors learn how to make these floral necklaces at Senator Fong’s plantation & gardens

37. Unite formally

38. Command post: Abbr.

39. Fill in the blank with this word: “___-Deutschland”

40. Govt.-issued funds

43. Lasting forever

44. Ravel’s “Gaspard de la ___”

46. Fill in the blank with this word: “”At the ___ Core,” 1976 sci-fi film”

47. Strong cotton fabric

48. Successor to Marshall on the Supreme Court

50. Whopper

51. Slangy moves

54. Zadora and Lindstrom

55. Poll amts.

56. Fill in the blank with this word: “___ buco”

57. Alley org.

58. Work started by London’s Philological Soc.

59. These 3 letters refer to a company’s liability, or an old Ford

60. Teenage hooligan, to a Brit

PUZZLE 20

ACROSS

1. Fill in the blank with this word: "___ point (makes sense)"

5. Not, in Nuremberg

10. UV blockage nos.

14. Put ___ in one's ear

15. Sargon II's god

16. Sci-fi princess

17. Much of Mongolia

18. Watts, Campbell, Judd

19. Mythical king of the Huns

20. Sharing of thoughts on a TV show?

23. Fill in the blank with this word: """If I ___ Rich Man"""

24. Fill in the blank with this word: "___-dieu"

25. Fill in the blank with this word: "___ Rancho (suburb of Albuquerque)"

26. What a patrol car might get, for short

27. Globe: Abbr.

30. Unpleasant smell

32. The English translation for the french word: estrade

34. Southeast Florida city

38. Paul Newman's directorial debut

42. A.L. home run champ of 1950 and '53

43. Haphazard

45. Primary strategy

48. Mike & ___ (candy brand)

50. Fill in the blank with this word: ""Tais-___!" (French "Shut up!")"

51. Fill in the blank with this word: ""Am ___ risk?"""

52. Fill in the blank with this word: "___ as a pin"

56. Treeless tract

58. Cheesy TV comedy?

62. Cub #21 of the 1990s-2000s

63. Stirs up

64. Fill in the blank with this word: "___-Jutsu (Japanese martial art)"

66. Graphic ___

67. Veins

68. Les …tats-___

69. Wild ___

70. Three-time Masters winner

71. Sandy tract, in Britain

DOWN

1. Witchy woman

2. Like a disappointing golf game

3. Growing post-W.W. II environs

4. Well-coordinated

5. Zola's streetwalker

6. This ___ outrage!'

7. Make like Pac-Man

8. Good place for a smoke

9. Singer Lopez

10. Verbal assault

11. Leader called "the Great"

12. Substitute

13. Caved in

21. Fill in the blank with this word: "___ in Charlie"

22. Fill in the blank with this word: "Cul-___"

23. Walletful

28. Wing: Prefix

29. Fill in the blank with this word: "___ brace (device used to immobilize the head and neck)"

31. Sch. staffer

33. The English translation for the french word: scannÈriser

35. Three ___ and One DJ' (Beastie Boys song)

36. Whose woods these ___ think…': Frost

37. What stripes may indicate

39. Midwest and Plains states, e.g.

40. Skip a dinner date

41. Zoo, so to speak

44. Fill in the blank with this word: ""O Sole ___""""

45. Log cutter

46. Site of Pakistan's Shalimar gardens

47. Some global treaty subjects, informally

49. Slippery ___

53. Noblemen

54. The Capris' "There's ___ Out Tonight"

55. Up on deck

57. Netanyahu's party

59. White House's ___ Room

60. Tiny tabby tormentor

61. Made a tax valuation: Abbr.

65. What's the ___?'

PUZZLE 21

1	2	3	4	■	5	6	7	8	■	9	10	11	12	13
14				■	15				■	16				
17				■	18				■	19				
20				21					22		■	23		
■	■	■	24				■	■	25		26			
27	28	29				■	30	31					■	■
32					■	33							34	35
36			■	■	37					■	■	38		
39			40	41					■	42	43			
■	■	44						■	45					
46	47					■	■	48				■	■	■
49			■	50		51	52					53	54	55
56			57		■	58				■	59			
60					■	61				■	62			
63					■	64				■	65			

ACROSS

1. Window's support

5. Palio di ___ (Italian horse race)

9. Porter's "Ev'ry Time ___ Goodbye"

14. Poulenc's "Sonata for ___ and Piano"

15. Take down ___ (humble)

16. You don't say it when you stand

17. Shakespeare's ___ of Salisbury

18. The English translation for the french word: loupe

19. First black major-league baseball coach Buck ___

20. Rest stops?

23. Walken's gift in "The Dead Zone"

24. Until 1990 it was the capital of West Germany

25. Zips

27. Royal address?

30. Hamlet's "___ and arrows"

32. Red Sox Hall-of-Famer Bobby

33. Like chicken fingers

36. Fill in the blank with this word: ""___ Wiedersehen"""

37. Fill in the blank with this word: “”It’s ___ against time””

38. Potus #34

39. Isolates

42. Take ___ (look)

44. Mouthed off

45. Turn

46. Obtrudes

48. Syrup brand

49. Maritime CIA

50. “Do tell!”

56. Washington bank name since 1840

58. Finish this popular saying: “While there’s life there’s_____.”

59. Tennyson’s “immemorial ___”

60. Not ___ (mediocre)

61. Record-setting miler Jim

62. The English translation for the french word: soyeux

63. Spanish beings

64. The U.N.’s Kofi ___ Annan

65. Je ne ___ quoi

DOWN

1. Average guys

2. Simple rhyme scheme

3. ‘Wandering at ___’ (Whitman poem)

4. Dwell on

5. Highly rated security

6. Turn back

7. Fill in the blank with this word: “___ Polo of “Meet the Fockers””

8. White house: Var.

9. Classic Abbott and Costello bit

10. Strauss’s “___ Heldenleben”

11. Like cliffs

12. Zipperless pants called broadfalls are worn by this U.S. group named for Jacob Ammann

13. Pound sounds

21. Unit of pressure

22. Wavelike design

26. Team V.I.P.: Abbr.

27. Fill in the blank with this word: “___ touch”

28. Fill in the blank with this word: “”___ lost me””

29. Privilege of liberals?

30. Big band era standard

31. Univ. worker

33. O’er bank and ___...he glanced away...’: Sir Walter Scott

34. Fill in the blank with this word: “”James Joyce” author Leon ___”

35. Slayton of Apollo 18

37. The English translation for the french word: apside

40. With 100-Across, Naples opera house Teatro di ___

41. Hockey stat

42. Fill in the blank with this word: “”___ Lee” (classic song)”

43. The English translation for the french word: professer

45. Connected series

46. Steed

47. Whopper topper

48. Shot

51. Servitude

52. Yankee Hall-of-Famer Waite ___

53. Inter ___

54. Year Otto I became king of the Lombards

55. Tut’s kin?

57. Stuff

PUZZLE 22

ACROSS

1. Tiny bit of a tick: Abbr.

5. This synonym for “kingdom” comes from the Latin for “regal”

10. Subj. of a 39-Down reminder

13. Fill in the blank with this word: “”All sales ___”””

15. Race of Norse gods

16. Put in one’s ___ (interfere)

17. Painful prod

19. Sault ___ Marie

20. Trollope’s “Lady ___”

21. Ancient city to which Paul wrote an Epistle

23. Finish this popular saying: “Make love not______.”

25. Tomcat

27. Looney Tunes character with a strong Southern accent

28. Roy Wood’s band before Wizzard

29. 1930s French premier Léon

31. T. ___

32. Heater stats

34. Yago Sant’___ (wine brand)

36. Sparked anew

40. Expansionist doctrine

43. Inflicted upon

44. Taking the maximum risk, as a poker player

45. War memento, maybe

46. Female rap trio with the #1 hit "Waterfalls"

48. River isles

50. The Carolinas' ___ Dee River

51. Support in a confessional

55. Staten Isl., for one

56. Popular film Web site, briefly

57. like or characteristic of a mirrored image

59. I.R.S. form 1099-___

61. When some stores open

62. Witch's hamper?

66. Rock's ___ Speedwagon

67. Thwart in court

68. Soixante minutes

69. Fill in the blank with this word: "___-80 (old computer)"

70. Leather strips that help a rider control a horse

71. Vast

DOWN

1. Producer: Abbr.

2. Sprechen ___ Deutsch?'

3. Evasive maneuver in football

4. "High Hopes" lyricist

5. Whack

6. Workplace fairness agcy

7. Want from

8. Try to avoid detection

9. Classic PBS name

10. Rapid, to Rossini

11. Volga region native

12. White House's ___ Room

14. Fill in the blank with this word: ""Harry Potter" character Neville ___"

18. Follow too closely

22. "Fiddler on the Roof" setting

23. Online health info site

24. Russia's ___ Republic

26. Anatomical sac

30. Uris's "___ Pass"

33. Tot tender

35. unscripted (similar term)

37. 1986 John Frankenheimer film, in Rome?

38. Gravestone abbr.

39. Relaxed personality

41. Copyist

42. Put away, in a way

47. Shade of red

49. -

51. The first one opened in Detroit in 1962

52. Gunpowder ingredient

53. Composer Dohn

54. Straight: Prefix

58. Rodenticide name

60. Go on a vacation tour

63. Tests for college credit, briefly

64. Verdi's "___ tu"

65. Shamus

PUZZLE 23

1	2	3	4	■	5	6	7	8	9	■	10	11	12	13
14				■	15					■	16			
17				18						■	19			
20					■	■	21			22				
■	■	■	23		24	25	■	■	26					
27	28	29					30	31				■	■	■
32			■	33					■	34		35	36	37
38			39	■	40				41	■	42			
43				44	■	45				46	■	47		
■	■	■	48		49						50			
51	52	53				■	■	54				■	■	■
55						56	57	■	■	58		59	60	61
62				■	63			64	65					
66				■	67					■	68			
69				■	70					■	71			

ACROSS

1. Work on the edge?

5. Made off with

10. Pics from which to make more pics

14. Town on the Thames

15. Fill in the blank with this word: "Easy ___"

16. When expected

17. Makeup applicator

19. To heat water to 212 degrees

20. Yawn

21. Thin skin

23. Shadow

26. Possible name for the first decade of the century

27. Roast the other side of the marshmallow?

32. Fill in the blank with this word: "___ Cayes, Haiti"

33. Fill in the blank with this word: ""All I ___ Do" (Sheryl Crow song)"

34. Indications

38. Combined, in Compi

40. Make final, as a deal

42. Workplace fairness agcy

43. Travel mag listing

45. Tagging along

47. Transfer ___

48. Hinge of a palindrome

51. Fill in the blank with this word: "___ oil (perfumery ingredient)"

54. Investment firm T. ___ Price

55. Name on many libraries

58. Wistful exclamation

62. Fill in the blank with this word: "___ prof."

63. Good neighbor policy

66. Quintillionth: Prefix

67. They thought C-3PO was a god in "Return of the Jedi"

68. Uncommon blood type, informally

69. One on a mission, maybe

70. Tried to tackle, say

71. Piped fuel

DOWN

1. Jazz buff

2. Texter's 'Alternatively ...'

3. Fill in the blank with this word: ""And ___ word from our sponsor""

4. Ultimately becomes

5. The English translation for the french word: secousse

6. Walken's gift in "The Dead Zone"

7. Fill in the blank with this word: """___ and away!"""

8. Fill in the blank with this word: """That's ___"""

9. Run again

10. "Don't sweat it"

11. Violinist/bandleader ___ Light

12. Fill in the blank with this word: "___ trip"

13. Monica ___, two-time U.S. Open champ

18. Form into an arch, old-style

22. San ___ Obispo, Calif.

24. Transfer and messenger materials

25. Work that includes a visit to the underworld

27. The English translation for the french word: g,cher

28. Watercolorist ___ Liu

29. Fill in the blank with this word: """Beauty ___ the eye Ö"""

30. Moroccan toppers: Var.

31. Overdrawn?

35. Frobe who played Goldfinger

36. Second to ___

37. The English translation for the french word: plaie

39. Oilers' home

41. You get weak-kneed fondling a tournament mallet used during chukkers in this sport

44. When food enters the small intestine, it's joined there by this fluid stored in the gallbladder

46. Victors' cry

49. Groundhog, notably

50. What is the capital of this country - Iran

51. Big tournaments for university teams, informally

52. Kodak founder

53. B. & O. stop: Abbr.

56. To stand in the center of this state, go 5 miles northeast of Ames & stand there--how exciting...

57. North Carolina university

59. ___-ho

60. Zeno of ___

61. Lines of thought, for short?

64. Longtime Chicago Bears coach

65. Fill in the blank with this word: """Humanum ___ errare"""

PUZZLE 24

1	2	3	4	5	■	6	7	8	9	■	10	11	12	13
14					■	15				■	16			
17					■	18				■	19			
20					21					22	■	23		
24			■	■	25				■	26	27			■
28			29	30		■	■	■	31					32
■	■	■	33			34	35	36		■	37			
38	39	40								41				
42				■	43							■	■	■
44				45		■	■	■	46			47	48	49
■	50				■	51	52	53		■	■	54		
55			■	56	57					58	59			
60			61	■	62				■	63				
64				■	65				■	66				
67				■	68				■	69				

ACROSS

1. Fill in the blank with this word: "___ a hatter"

6. Legal scholar Guinier

10. The English translation for the french word: ‡mha

14. Vienna State Opera music director starting in 2002

15. Those: Sp.

16. North Carolina's Cape ___

17. Popular A.M. host

18. Spittoon sound

19. Fill in the blank with this word: "___-Alt-Del"

20. Controlling things once more

23. Fill in the blank with this word: ""Some ___ meat and canna eat": Burns"

24. GM: "___ the USA in your Chevrolet"

25. Runs a tab

26. Oft-quoted auth.

28. The English translation for the french word: tatami

31. Singer Twain

33. Poppycock

37. Washington chopping down the cherry tree, e.g.

38. Bishop preparing to hold a yard sale?

42. Batting helmet feature

43. Whatnot

44. The English translation for the french word: sillon

46. Time-___

50. Wild Indonesian bovine

51. Fill in the blank with this word: “”Young ___ Boone” (short-lived 1970s TV series)”

54. Slo-___ fuse

55. Fill in the blank with this word: “Circular ___”

56. ...

60. The Bard of ___ (Shakespeare)

62. Military org. with the motto “Per ardua ad astra”

63. Saxophonist ___ Lawrence

64. Bookstore sect.

65. Vapory beginning

66. Fill in the blank with this word: “___ Park, Calif.”

67. “As we have therefore opportunity, let ___ good to all men”: Galatians

68. Plaintiff

69. TV newsman David

DOWN

1. Really bother

2. Wild honeysuckle, e.g.

3. French writer Alphonse or L

4. Wowed

5. Hall & Oates: “_____ Smile”

6. Pepe ___

7. Fill in the blank with this word: “”___ World Turns””

8. Voyagers: Suffix

9. Fill in the blank with this word: “”Beauty ___ the eye Ö””

10. Portlandia’ network

11. Antifreeze ingredient

12. Cio-Cio-San’s way out

13. The English translation for the french word: orle

21. Well-groomed

22. Suffix with bull or bear

27. Saintly circles

29. Sound before a blessing

30. Kenyan president Daniel arap ___

31. Places such as Anatevka in “Fiddler on the Roof”

32. Truman’s nuclear agcy.

34. Lb. or oz.

35. Obama’s signature health law, for short

36. Wilt

38. The English translation for the french word: simbleau

39. Uncommon sort

40. Tree with very hard timber

41. Realm of Otto I: Abbr.

45. Winery sight

47. Strip alternatives

48. Winter 1997-98 newsmaker

49. Turndown?

51. Mirabile ___ (wonderful to say)

52. What’s in ___?’

53. See 41-Down

55. He played Mowgli in “Jungle Book”

57. Round dances

58. Fill in the blank with this word: “”...gimble in the ___”: Carroll”

59. 1960 Updike novel

61. Vietnam’s ___ Dinh Diem

PUZZLE 25

1	2	3	4	■	5	6	7	8	9	■	10	11	12	13
14				■	15					■	16			
17				18						19				
20								■	■	21				
■	■	■	■	22			■	23	24			■	■	■
25	26	27	28				29					30	31	32
33					■	■	34				■	35		
36				■	37	38				■	39			
40			■	41				■	■	42				
43			44					45	46					
■	■	■	47				■	48			■	■	■	■
49	50	51			■	■	52				53	54	55	56
57					58	59								
60				■	61					■	62			
63				■	64					■	65			

ACROSS

1. The English translation for the french word: oued

5. Violinist Zimbalist

10. Waste

14. toward the mouth or oral region

15. Game, to Guglielmo

16. Fill in the blank with this word: "___ limits (election issue)"

17. Huey

20. Impression of Count Dracula?

21. Repeat calls?

22. Fill in the blank with this word: "Carolina ___"

23. Summer Games org.

25. Is pessimistic

33. P.M. between Netanyahu and Sharon

34. Sea eagles

35. Obama's signature health law, for short

36. Sartre's "No ___"

37. Trinity member

39. Location

40. Fill in the blank with this word: "Ad

___ (relevant)"

41. Baloney

42. Feudal lands

43. Assassinated leader called "the Liberator"

47. Omnia vincit ___

48. Fill in the blank with this word: "Bon ___"

49. Yemeni's neighbor

52. Sexy movie companions, maybe

57. Not so much

60. Prefix with -meter

61. Middle: Prefix

62. Fill in the blank with this word: "Feel the ___"

63. Worry

64. Exceptional rating

65. Strong fiber

DOWN

1. Woman-chaser

2. Gray ___

3. Vampire's curfew

4. Yeah, man!'

5. The English translation for the french word: ÈgoÔste

6. R

7. Weapon in the game Clue

8. Umberto ___, author of 'The Name of the Rose'

9. Unruly head of hair

10. The name of this plaster used in the 18th C. to make faux marble refers to rough exterior coating

11. Grateful Dead bassist Phil

12. Singer India.___

13. Yukons, e.g.

18. Defeat

19. Tom's cries

23. Pulitzer winner for "Driving Miss Daisy"

24. Hong Kong's Hang ___ Index

25. Movie critic Roger

26. Some horizontal lines

27. Fill in the blank with this word: "___ facie"

28. Finish this popular saying: "You are what you_____."

29. The English translation for the french word: tÈlex

30. Autumn toiler

31. Wells works

32. Tibiae neighbors

37. Hawaii's ___ Bay

38. Walkie-talkie word

39. '___ you to horse': Macbeth

41. Fill in the blank with this word: "___ Oliver, a k a the Naked Chef"

42. Cremona violin

44. Was nearly out

45. Loan payment schedules: Abbr.

46. Observant ones

49. Wagering sites, for short

50. Water under the bridge

51. Fill in the blank with this word: "Europe's Gorge of the ___"

52. Nesters

53. Three men in ___'

54. Trillion: Prefix

55. Some specialize in elec.

56. Put back

58. Barges

59. Le coeur a ___ raisons...': Pascal

PUZZLE 26

ACROSS

1. Take ___ at (try)

6. Omar of “The Mod Squad,” 1999

10. UV blockage nos.

14. State in Brazil

15. Soft roe

16. Classic Nestl

17. Fill in the blank with this word: “___ Line (German/Polish border)”

19. Sell short

20. The English translation for the french word: RAU

21. What’s tapped at a beer bust

22. Violent, perhaps

24. Some brass

28. Fill in the blank with this word: “”___ beam up” (“Star Trek” order)”

30. Purpose

31. Knuckler alternative

32. This one’s ___’

33. Prefix with fauna

36. Whiskies

37. Tragic James Fenimore Cooper

character

39. Place for keys and lipstick

40. Unrealized 60's Boeing project

41. Summer Games org.

42. Brewer Adolphus

43. Smashes from Sampras

45. Where ___

46. *One who's often doing favors

50. Writers Shreve and Brookner

51. Military asst.

52. Barges

55. Pro ___ (proportionately)

56. What it's like to be Spider-Man?

60. Many a holiday visitor / Bandit

61. Webzine

62. Trig angle symbol

63. Fill in the blank with this word: "___ Vista"

64. Trails off

65. Fill in the blank with this word: ""Super Duper ___" (anime series)"

DOWN

1. Leigh Hunt's "___ Ben Adhem"

2. Thompson of "Family"

3. having the property of becoming permanently hard and rigid when heated or cured

4. You can bob for apples because they're 25% this, which allows them to float

5. Withdrawal carrying a steep penalty?

6. Solzhenitsyn, e.g.

7. TV's Magnum and others

8. Some nouns: Abbr.

9. Breastbones

10. Plaza

11. Fill in the blank with this word: "___ Arenas, port in 93-Down"

12. Cuba's Castro

13. Variety listings

18. Want ad inits.

23. Adjust, as a clock

25. Lawyers: Abbr.

26. Trattoria dumplings

27. Writer ___ St. Vincent Millay

28. Bank-to-bank transactions: Abbr.

29. House Committee on ___ and Means

33. Auto-stopping innovation

34. Fill in the blank with this word: "Explorer Cabeza de ___"

35. See 103-Across

37. Assessment paid only by those who benefit

38. Norse goddess of fate

39. Tutsi foe

41. Part of the eye

42. Pumas, e.g.

43. What is the capital of this country - Canada

44. Fill in the blank with this word: ""Lowdown" singer Boz ___"

46. Santa ___

47. With everything counted

48. Join, as a table

49. When some summer reruns are broadcast: Abbr.

53. Fill in the blank with this word: "Baseball's ___ Gaston"

54. Flies away

57. Pres. appointee

58. Snatch

59. Moo ___ pork

PUZZLE 27

1	2	3	4	■	5	6	7	8	9	■	10	11	12	13
14				■	15					■	16			
17				18						■	19			
20			■	21						■	22			
23			24			■	■	■	25	26			■	■
■	■	■	27			28	29	30					31	32
33	34	35			■	36					■	37		
38				■	39					■	40			
41			■	42					■	43				
44			45						46			■	■	■
■	■	47				■	■	■	48			49	50	51
52	53			■	54	55	56	57			■	58		
59				■	60						61			
62				■	63					■	64			
65				■	66					■	67			

ACROSS

1. Sony subsidiary

5. an ethnic group living in Azerbaijan

10. TV's 'How ___ Your Mother'

14. Torre and Valentine: Abbr.

15. Puerto ___

16. Fill in the blank with this word: "Big ___"

17. Makeup applicator

19. Tire, at the Michelin plant

20. Mail Boxes ___

21. Thistlelike plant

22. When Parisians typically vacate

23. Word seen twice on a U.S. map

25. With a 2007 women's water polo title, this California school became the first to win 100 NCAA titles

27. Event that includes Snowboarding Charades and Motocross 20 Questions?

33. They're S-shaped

36. Japanese immigrant

37. Singer ___ Rose

38. Fill in the blank with this word: "___ diagram"

39. Several Norwegian kings

40. Madrid month

41. Treebeard, e.g.

42. Fill in the blank with this word: “”Thy word is ___ unto...”: Psalms”

43. Corporate shuffling, for short

44. Close call

47. Spontaneous skits

48. Lacking radiating appendages, as nerve cells

52. The English translation for the french word: gala

54. Slip by

58. Work without ___

59. Touched the tarmac

60. Reagan Administration figure

62. Oscar-winning French film director ___ Cl

63. Time waster

64. Wildcats' org.

65. Officially listed: Abbr.

66. They test reasoning skills: Abbr.

67. Wooley with the 1958 #1 hit “The Purple People Eater”

DOWN

1. Pumped

2. Jim Croce's “___ Name”

3. Second-largest city of Rhode Island

4. Comparable in years

5. Unpaid debt

6. Fill in the blank with this word: “”___-Dee-Doo-Dah””

7. Old French coins

8. Fill in the blank with this word: “___ McCawley, Ben Affleck's role in “Pearl Harbor””

9. Streams of arrivals

10. The English translation for the french word: impala

11. Without assistance in a fight

12. Flightless bird: Var.

13. Unbending

18. Singers James and Jones

24. U.S. ___ (annual Queens event)

26. The English translation for the french word: CGI

28. Philanthropist Wallace and others

29. Target of a decade-long manhunt, informally

30. Wedding count

31. Swimming laps, e.g.: Abbr.

32. Trudge

33. Fill in the blank with this word: “Dutch ___”

34. Fill in the blank with this word: “___ splicing”

35. Making necessary

39. What Vito Corleone's company imported

40. Fill in the blank with this word: “___ Aarnio, innovative furniture designer”

42. Without further ___

43. Write again

45. Violent, perhaps

46. Stunning weapons

49. Tough, durable wood

50. Temple architectural features

51. Where ‘they tried to make me go,’ in an Amy Winehouse hit

52. Teri ___, Best Supporting Actress nominee for ‘Tootsie’

53. You'll use up 3 vowels playing this word that means toward the side of a ship that's sheltered from the wind

55. Some legal scholars, for short

56. Teatro ___ Scala

57. Painter Mondrian

61. Switch ups?

PUZZLE 28

ACROSS

1. O.R. doings

5. Finish this popular saying: "Every stick has two_____."

9. Manhandle

14. Fill in the blank with this word: "___ Minor"

15. Sub ___ (secretly)

16. visible (similar term)

17. More of the answer

19. Weird

20. Young raptor

21. Resident of Asmara

23. Groove-billed ___

24. Exceptional rating

25. Root beer float with chocolate ice cream

29. Immune system lymphocytes

33. Get an ___ (ace)

34. There are lots of eaters of chocolate bunnies on this holiday

36. How to ___ knot (Boy Scout's lesson)

37. Work in a refinery

39. Chem. formula for hydrogen isocyanide

40. Jim-dandy

41. Star of "Mr. Hulot's Holiday"

42. Tiled art

44. Fill in the blank with this word: ""Humanum ___ errare""

45. Social breakdown

47. Repeated musical phrase

49. Nevil Shute's '___ Like Alice'

51. Weapons prog. since 1983

52. Trattoria order

55. Ultimatum word

59. What road hogs hog

60. "Tomb Raider" adventuress

63. Possible answers for 20-Across

64. Sinclair Lewis's ___ Timberlane

65. Princess loved by Hercules

66. Fill in the blank with this word: "___ New Guinea"

67. Like some high-fiber cereal

68. Their days are numbered

DOWN

1. Word processor command

2. Voice of America org.

3. Unscramble this word: rnig

4. Stop-press order?

5. Works with steam?

6. Emergency call

7. The English translation for the french word: DSM

8. MS. enclosures

9. Westminster Abbey area

10. Resisting

11. Fill in the blank with this word: ""___ #1!""

12. The English translation for the french word: mÈlodie

13. Kind of list

18. Milk curdler

22. They believe

24. Housekeeper player on "Benson"

25. July 4th concert site

26. Times ___

27. You're __ talk!'

28. Vitellius succeeded him

30. Fill in the blank with this word: ""___ Rock" (Bob Seger hit)"

31. Unscramble this word: stela

32. Tell

35. They were once "The most trusted name in television"

38. Capital originally called the City of the Kings

40. Vinegary

42. Fill in the blank with this word: ""Love ___ leave ...""

43. "Piece of cake!"

46. Lake ___, source of the Mississippi

48. Fill in the blank with this word: "___ acid"

50. Radiotelephony response

52. Thunder sound

53. Rubaiyat' rhyme scheme

54. The L train?

56. Hula ___

57. Black key

58. Ways: Abbr.

61. Fill in the blank with this word: ""___ approved" (motel sign)"

62. What's right in front of U

PUZZLE 29

1	2	3	4	5	■	6	7	8	9	10	■	11	12	13
14					■	15					■	16		
17					18						■	19		
20						■	21				22			
23			■	24		25	■	■	26				■	■
■	■	■	27				28	29	■	30			31	32
33	34	35		■	36				37	■	38			
39				■	40					■	41			
42				■	43					■	44			
45				46	■	47				48		■	■	■
■	■	49			50	■	■	51			■	52	53	54
55	56					57	58	■	59		60			
61			■	62				63						
64			■	65					■	66				
67			■	68					■	69				

ACROSS

1. Spy Rudolf and others

6. Lobster feelers

11. Fill in the blank with this word: "1968 hit "Harper Valley ___""

14. Photo finish?

15. William Browne poem "___, as Fair as Ever Saw the North"

16. TV band above channel 13, in brief

17. Match for a bad guy?

19. Stretch out, in Scotland

20. Peasants' cooperatives

21. Whitewall, maybe

23. Wind dir.

24. Young fellow

26. St. John's ___ (herbal remedy)

27. Strip's cry of disgust

30. Spiral-horned African antelope

33. www page creation tool

36. Works with

38. Use tiny scissors

39. Cries of surprise

40. Fill in the blank with this word: "___, meenie, miney, mo"

41. Uses a riflescope

42. Symphonie espagnole' composer

43. Abbr. after Ted Kennedy's name

44. Summer cooler

45. Weenie

47. Equal in height

49. Hic, ___, hoc

51. Fill in the blank with this word: ""An' singin there, an' dancin here, / Wi' great and ___": Burns"

52. Pou ___ (vantage point)

55. Something in need of change

59. String substitute?

61. Thieves' take

62. *Stale air removers

64. World production of this is now 84 million barrels a day, up about 10 million barrels from 1997

65. Tube

66. Famed statement by 67-Down

67. Fill in the blank with this word: "___ Ronald Reagan"

68. This garlic-flavored mayonnaise from Provence is popularly served with fish

69. Wasp homes

DOWN

1. Turkish pooh-bahs

2. What "Britney Spears" means in rhyming slang

3. Related through the mother

4. Unscramble this word: ivle

5. Strait of Messina menace

6. This word for a buddy or chum is often followed by "around"

7. Fill in the blank with this word: "Bust ___ (laugh hard)"

8. Norse fire god

9. Yeah, right!'

10. Places atop

11. Overly strict

12. Fill in the blank with this word: """___ she blows!"""

13. Fire

18. Sigmoid

22. Aid for upwind maneuvers

25. Woman's name meaning "peace"

27. Loser

28. Lollobrigida and others

29. Unauthorized withdrawals?

31. Whitewash

32. Vaulted space

33. Weekend Today' anchor Lester

34. Unfreeze

35. Minor problems, so to speak

37. Some oilseeds

46. Una ___ (old coin words)

48. formerly used for church utensils

50. Year in Marcus Aurelius's reign

52. Trash-talks

53. Touch of color

54. Tobacco kilns

55. Dynasty in which Confucianism and Taoism emerged

56. Votes overseas

57. Fill in the blank with this word: """___ that's your game!"""

58. Old Chinese money

60. Without ___ (pro bono)

63. Fill in the blank with this word: "___ possidetis (as you possess, at law)"

PUZZLE 30

1	2	3	4	5	6		7	8	9	10		11	12	13
14							15					16		
17						18						19		
20						21						22		
23					24						25			
			26	27					28					
29	30	31				32	33	34			35			36
37				38					39	40		41		
42			43			44					45			
	46			47	48				49					
50						51	52	53			54	55	56	57
58				59						60				
61				62					63					
64				65					66					
67				68					69					

ACROSS

1. Together

7. Youngsters

11. Union ___: Abbr.

14. Response to 'I have to be going now'

15. Take ___ (swing hard)

16. Test result, at times: Abbr

17. Show featuring the scheming Dr. Zachary Smith

19. Visual way to communicate: Abbr.

20. Unscramble this word: aetss

21. Volleyball kill

22. Year the emperor Decius was born

23. Fill in the blank with this word: "___ Verde National Park"

24. Weatherproof headlight

26. Place runners?: Abbr.

28. Colloquial possessive

29. Ladle cradle

35. Fill in the blank with this word: """___ oui!"""

37. 'My mama done ___ me ...'

38. Singer with the 1975 #1 hit "Lady

Marmalade"

41. Brain and spinal cord: Abbr.

42. Salon creation

44. Definition, part 2

46. Fill in the blank with this word: "Beethoven's "Concerto No. 5 in ___ major"""

49. Potus #34

50. Relieved

54. Fill in the blank with this word: "___ Bora, wild part of Afghanistan"

58. Untilled tract

59. Mystery writer Gardner et al.

60. Writing by Montaigne

61. Actor Herbert

62. Threatens violence

64. Fill in the blank with this word: "Bad ___, Mich. (seat of Huron County)"

65. Explosives

66. Smelly smoke

67. Old what's-___-name

68. Fill in the blank with this word: "Dionne Warwick's "I ___ Little Prayer"""

69. Steamy, say

DOWN

1. Nation of ___

2. Lasso loop

3. Exceptional rating

4. Spanish valentine sentiment

5. Fill in the blank with this word: """___ can't be!"""

6. This is ___'

7. Phi Beta ___

8. Train track beam

9. Detachable shirtfront

10. Swimwear brand

11. Sputnik 1 launched it

12. Longtime NBC Symphony conductor

13. Fill in the blank with this word: """There's ___ chance of that"""

18. Wind dir.

24. Purchase from a jeweler

25. The English translation for the french word: vagabond

27. Anguilla is part of it: Abbr.

29. Fill in the blank with this word: "Disco ___ (character on "The Simpsons")"

30. Pontiff for just 26 days in 1605

31. Exes, of a sort

32. Withdraw

33. GM: "___ the USA in your Chevrolet"

34. Female rap trio with the #1 hit "Waterfalls"

36. Unrealized 60's Boeing project

39. "This Gun for Hire" star

40. Fix a squeak

43. Grand ___ Opry

45. Obsolete court tactic

47. Turns away

48. European capital

50. To whom a Muslim prays

51. See 23-Across

52. Fill in the blank with this word: "2010 Olympic ice dancing gold medalist ___ Virtue"

53. Yacht's dir.

55. Fill in the blank with this word: "___ orange"

56. Bad-tasting

57. Support person

60. I could ___ horse!'

63. Ukraine, e.g., formerly: Abbr.

PUZZLE 31

1	2	3	■	4	5	6	7	8	■	9	10	11	12	13
14			■	15					■	16				
17			■	18					19					
20			21		■	22						■	■	■
23					24		■	25			■	26	27	28
■	■	■	29				30	■	31		32			
33	34	35		■	36			37	■	38				
39				40					41					
42					■	43				■	44			
45					46	■	47			48		■	■	■
49			■	50		51	■	52				53	54	55
■	■	■	56				57		■	58				
59	60	61							62		■	63		
64					■	65					■	66		
67					■	68					■	69		

ACROSS

1. What's funded by FICA, for short

4. What the Laugh Factory produces

9. Wind through Darwin, Minnesota or Cawker City, Kansas; they both claim the world's largest ball of this

14. This ancient Hebrew measure equal to about 2 quarts sounds like a synonym for "taxi"

15. Fill in the blank with this word: "Allan-___, Robin Hood companion"

16. Rowed

17. So that's ___?'

18. Romantic dinner reservation

20. It may be thrown in the ring

22. The English translation for the french word: incorporation

23. 'Hoarders' subject

25. Volga feeder

26. Prodigal ___

29. Israeli seaport

31. "Bewitched" witch

33. Wild guess

36. Waist circlers

38. View through the crosshairs

39. Himalayan aviation board?

42. Prefix meaning "likeness"

43. Oscar Wilde poem "The Garden of ___"

44. Tupperware stock

45. Fill in the blank with this word: ""Pretty ___""

47. With it

49. Was on the bottom?

50. Yukon S.U.V. maker

52. Fatigued

56. See 34-Across

58. Where ___

59. Know-it-all

63. Fill in the blank with this word: "___ Maria"

64. Whoop

65. Johnny ___, "Key Largo" gangster

66. Scottish Peace Nobelist John Boyd ___

67. Old Apple computers

68. The results ___'

69. Verdi's "___ giardin del bello"

DOWN

1. The English translation for the french word: scalp

2. Yemen's capital

3. Possible answers for 20-Across

4. John Wayne film set in Africa

5. Writer ___ Louise Huxtable

6. Fit to be lived in

7. Taking the maximum risk, as a poker player

8. Tend

9. Was out for the afternoon?

10. Tech expert, as it were

11. Fill in the blank with this word: ""Just Another Girl on the ___" (1993 drama)"

12. Like ___

13. Tokyo, once

19. Sham

21. Japanese flower-arranging art

24. Crooked

26. Sleep: Prefix

27. Waxed bombastic

28. Big-time competition: Abbr.

30. The name of this jeweled coronet worn by princesses is from the Greek for "turban"

32. The English translation for the french word: dialecte

33. Sounds that may be heard before bangs?

34. Santa ___, El Salvador

35. Fill in the blank with this word: ""You're ___ and don't even know it""

37. Guard against drifting

40. Silicon Valley city

41. Those: Sp.

46. Peggotty girl in "David Copperfield"

48. Fill in the blank with this word: "___ d'etre"

51. You Can't Take It With You' director

53. Capital of Jamaica [black]

54. New ___ (Congolese money)

55. Video game pioneer

56. Swift Malay boat

57. Openness

59. Word part: Abbr.

60. Fill in the blank with this word: "___ tai"

61. Words to masseuses

62. Cable TV giant

PUZZLE 32

ACROSS

1. W.W. II vessel: Abbr.

4. Fill in the blank with this word: “Banda ___ (2004 tsunami site)”

8. Telephone part

14. 1923 earthquake site

16. Fill another teacup

17. Magnificent

18. Wife of King Mark of Cornwall

19. Newspaper's ___ page

20. Power ___

22. The English translation for the french word: complÈter

23. Essential amino acid

25. The English translation for the french word: b,tonner

27. Word of contempt

29. Literary character whose name is said to mean “laughing water”

32. Stonecrop

35. Fill in the blank with this word: “”___-daisy!””

37. Undergoes

38. Tom ___, 1962 A.L. Rookie of the Year

39. Fill in the blank with this word: "China's Chairman ___"

40. Thermonuclear experiment of the '50s

42. Spanish queen until 1931

43. Transcaucasian capital

44. Vital carrier

45. Swedish soprano noted for her Wagnerian roles

48. Rating of a program blocked by a V-chip

50. Comes to

52. Fill in the blank with this word: """___, she's mine ..." (Manfred Mann lyric)"

56. Certain constrictor

58. Expression of disgust

60. Wife of Esau

61. Wise one

63. Criticize

65. Rates of return

66. High card up one's sleeve

67. Italian dancer ___ Cecchetti

68. Lines of thought, for short?

69. Stock units: Abbr.

DOWN

1. Popular disinfectant

2. Soaked

3. Bell sounds

4. Philip of 'Kung Fu'

5. Mooches

6. The ruler of Qatar is known by this 4-letter title

7. Was artificially cooled, for short

8. Fill in the blank with this word: "Dernier ___"

9. With a hyphen added, this 6-letter word meaning "to quit" can mean "to join again", as in pro sports

10. Cave dwellers

11. Stadium special

12. Whopper

13. Ziegfeld Follies designer

15. Ismene's father

21. Puts in a blue funk

24. Fill in the blank with this word: """The Secret of ___" (1982 film)"

26. Touch-and-go

28. From a personal standpoint

30. Fill in the blank with this word: """And ___ thou slain the Jabberwock?"""

31. Wire-haired terrier of film

32. World War II weapon

33. About

34. Redistributed

36. U.P.S. delivery

40. Soccer star Mia

41. Right in every detail

43. Fill in the blank with this word: "Feather ___"

46. Mogadishu resident

47. Seinfeld, for one

49. One on a longship

51. Wake Up Little ___' (#1 Everly Brothers hit)

53. Writer Buchanan and others

54. Toothed bar

55. They may be tapped for the stage

56. Fill in the blank with this word: """For ___ sow..."""

57. ___ Capital (firm co-founded by Mitt Romney)

59. Walking stick

62. Taiping Rebellion general

64. Fill in the blank with this word: "___ Cayes, Haiti"

PUZZLE 33

1	2	3	4		5	6	7	8	9		10	11	12	13
14					15						16			
17				18						19				
20						21				22				
			23		24				25					
26	27	28					29	30						
31							32					33	34	35
36					37	38					39			
40				41					42	43				
			44					45						
46	47	48						49						
50						51	52			53		54	55	56
57					58				59					
60					61						62			
63					64						65			

ACROSS

1. Police dept. employee

5. The English translation for the french word: pincer

10. Old Chinese money

14. Fill in the blank with this word: “Dragon's ___ (early video game)”

15. Fill in the blank with this word: “2003 Nick Lachey hit “___ Swear””

16. Fill in the blank with this word: “”___ kleine Nachtmusik””

17. Cashier's error, as suggested by 17-, 22-, 47- and 58-Across?

20. Yawning or visibly astonished

21. Japanese prime minister Taro ___

22. Shoot-'em-up

23. Unwarm welcome

25. In a risqu√© fashion

26. *Destitution

31. Morning glories

32. Melville's Billy

33. See 20-Across

36. Wriggling

37. Fill in the blank with this word: “___

fixes"

39. Realtor's specialty, for short

40. O-___-O (brand of sponge)

41. Treaty of Nanking port

42. Fill in the blank with this word: "___ Beanie Babies"

44. "Dallas" spinoff

46. Best-selling novelist about whom Gore Vidal said "She doesn't write, she types!"

49. Mahler's "Das Lied von der ___"

50. Keep ___ out for

51. Worrying sound to a balloonist

53. Thompson and Watson

57. Memo about Stephen King's "Christine"?

60. Vissi d'___' (Puccini aria)

61. One in a line

62. Compass points (seen spelled out in 20-, 26-, 43- and 53-Across)

63. W.W. I plane

64. Gianni's grandmother

65. Reps.' rivals

DOWN

1. Russian gold medalist ___ Kulik

2. Fill in the blank with this word: "Da ___, Vietnam"

3. Fill in the blank with this word: "Director Vittorio De ___"

4. Predict

5. Series

6. Silent character in "Little Orphan Annie"

7. Wrestler Flair and others

8. Fill in the blank with this word: ""Lord, it is good for ___ be here" (words of Peter to Jesus)"

9. Open ___ night

10. Precursor to a historical "party"

11. ___ a stinker?' (Bugs Bunny catchphrase)

12. TV actress Georgia

13. The English translation for the french word: voir díun mauvais úil

18. Tend to again, as an injured joint

19. Stockpile

24. One of the Apostles

25. Symbol of Communism

26. Wait ___!' ('Hold on!')

27. One of the housewives on "Desperate Housewives"

28. Well-intentioned girl of rhyme?

29. Toes the line

30. Somme sight

33. Fill in the blank with this word: """___, vidi, vici"""

34. Fill in the blank with this word: "___ Nordegren, ex-wife of Tiger Woods"

35. Film director Nicolas

38. On the ___

39. Petition for again

41. Queen ___ War

43. Wrapped up

44. Walloped but good

45. Workbook unit

46. Writers Paretsky and Ryan

47. Union requirement, maybe?

48. Philadelphia train system

51. Like some vino

52. Takes root

54. What writer's block may block

55. Up and ___!'

56. Propagates

58. W-2 info: Abbr.

59. Spanish queen until 1931

PUZZLE 34

1	2	3	4	■	5	6	7	8	■	9	10	11	12	13
14				■	15				■	16				
17				■	18				19					
20				21		■	■	22			■	23		
24						25	26				27	■	■	■
■	28				■	29			■	■	30	31	32	33
34			■	35	36			■	37	38				
39			40					41						
42						■	43				■	44		
45				■	■	46			■	47	48			■
■	■	■	49	50	51				52					53
54	55	56	■	57			■	■	58					
59			60				61	62		■	63			
64					■	65				■	66			
67					■	68				■	69			

ACROSS

1. French author ___ Prevost

5. Tamiroff of "For Whom the Bell Tolls"

9. Not so well

14. What is the capital of this country - Switzerland

15. Fill in the blank with this word: "___ Linda, Calif."

16. To voluntarily give up a claim or a right, maybe with a gesture meaning bye-bye

17. Fill in the blank with this word: "___ as a doornail"

18. Captain Bligh after the mutiny?

20. Fill in the blank with this word: "Author ___ Le Guin"

22. Trigonometry abbr.

23. Uno, due, ___

24. Go ballistic

28. Wreck-checking org.

29. Flash ___ (faddish assembly)

30. The English translation for the french word: imam

34. Week-___-glance calendar

35. Take ___ of (sample)

37. Group with a board of governors

39. #1 movie of 1985

42. Regular care

43. Fill in the blank with this word: "___-Tass news agency"

44. Seoul soldier

45. Very recently: Abbr.

46. Subj. that deals with mixed feelings

47. 1940's-50's All-Star Johnny

49. "Take my word for it"

54. Fill in the blank with this word: "Actress ___ Scala"

57. Hall-of-Fame basketball coach Hank

58. Victorian, in a way

59. Baker Street group

63. Sci. course

64. Open shot

65. Start of the 22nd century

66. Leslie Caron musical

67. The English translation for the french word: insuffisant

68. Voice of Israel' author

69. Writer's supply: Abbr.

DOWN

1. Kareem ___-Jabbar

2. Result of a video viewer's spill?

3. The basics

4. Transfuses

5. Fill in the blank with this word: ""Wozzeck" composer ___ Berg"

6. Hootchy-___

7. The Monkees' '___ Believer'

8. Miniature auto brand

9. James Taylor's "___ Fool to Care"

10. My ___, Vietnam

11. Vocal rise and fall

12. Fill in the blank with this word: "___ and anon"

13. Midnight alarm giver

19. Shad ___

21. Like some oak leaves

25. Vent, in a way

26. someone whose reasoning is subtle and often specious

27. Wordsmith's ref.

31. Its capital is Nouakchott

32. 1993 Robert De Niro film

33. The English translation for the french word: humble

34. What ___!' ('That price is great!')

36. Sponge (up)

37. Painter's deg.

38. Screen blinker

40. Fill in the blank with this word: ""Etta ___""

41. Part of a dict. entry

46. It begins "In the Lord I take refuge"

48. Year-by-year accounts

50. Finish this popular saying: "Two wrongs don't make a_____."

51. Fill in the blank with this word: ""___ Roi" (Alfred Jarry play)"

52. Rose-red dye

53. Tibetan legends

54. Stored computer images, for short

55. High-performance Camaro ___-Z

56. Pianist Claudio

60. They dug his grave ___ where he lay': Sir Walter Scott

61. Pres. appointee

62. New York's former ___ Building

PUZZLE 35

1	2	3	4	■	5	6	7	8	9	■	10	11	12	13
14				■	15					■	16			
17				18						■	19			
20					■	21			■	22				
23			■	24	25				26			■	■	■
■	■	■	27				■	28				29	30	31
32	33	34				■	35							
36				■	37	38				■	39			
40				41				■	42	43				
44							■	45				■	■	■
■	■	■	46				47				■	48	49	50
51	52	53			■	54			■	55	56			
57				■	58				59					
60				■	61					■	62			
63				■	64					■	65			

ACROSS

1. ___ room (site of postdebate political commentary)

5. Fill in the blank with this word: "___ dark space (region in a vacuum tube)"

10. My Heart Can't Take ___ More' (1963 Supremes song)

14. The muse of history

15. Fill in the blank with this word: "___ as a rock"

16. Tampico track transport

17. Symbol of virility

19. Fill in the blank with this word: "Chris ___, 1988 N.L. Rookie of the Year"

20. Vantage point of Zeus, in Homer

21. One of the Cyclades

22. Villainous Shakespearean roles

23. Fill in the blank with this word: ""___ get it!" ("Aha!")"

24. Surfing gear

27. Fill in the blank with this word: ""___ for the poor""

28. Orioles owner Peter

32. Emergency situation

35. 14-Across's role, often

36. Fill in the blank with this word: ""Time ___ a premium""

37. Wigwam relative

39. Unscramble this word: romo

40. Plummet

42. Fill in the blank with this word: "___ Pieces"

44. Meadowsweet

45. Wrapper abbr

46. Garden perennials

48. Work ___ sweat

51. Jazz vocalist Carmen ___

54. Verse starter?

55. Spoilers?

57. Fill in the blank with this word: "Et ___"

58. Streamliner segment

60. Part of USPS: Abbr.

61. Wide collars

62. Where the laity sits

63. Mr. ___ of "Peter Pan"

64. Is stranded

65. Like two peas in ___

DOWN

1. Stupid jerk

2. Sylvia who wrote "The Bell Jar"

3. Five, on a gunslinger's gun

4. L'...toile du ___, Minnesota's motto

5. Camera operator's org.

6. Hallmark of a perfect game

7. Vegetable fats

8. Opposite of encourage

9. When some summer reruns are broadcast: Abbr.

10. "Shhhh!" follower

11. "King Lear" or "Hamlet": Abbr.

12. Biblical peak

13. Lennon's in-laws

18. Two-masters

22. Just let ___'

25. Less popular, as a restaurant

26. Transposes

27. That, to this

29. Where you might be among Hmong

30. Western Indian

31. Theological schools: Abbr.

32. Cookie containers

33. Fill in the blank with this word: ""Give ___ to Cerberus" (Greek and Roman saying)"

34. Old capital of Romania

35. Ursine : bear :: pithecan : ___

38. Clear out, as before a hurricane

41. Fill in the blank with this word: "Christine ___, "The Phantom of the Opera" heroine"

43. "Dallas" family name

45. You'll need six of them to finish this puzzle

47. Reserved bar?

48. Take the top off

49. Fill in the blank with this word: "___ Nurmi, the Flying Finn"

50. Comparable to a rose?

51. Unscramble this word: msas

52. Skelton's Kadiddlehopper

53. To laugh, to Lafayette

56. Insurance giant

58. What you might do with hat in hand

59. Visual way to communicate: Abbr.

PUZZLE 36

1	2	3	4	■	5	6	7	8	9	■	10	11	12	13
14				■	15					■	16			
17				18						■	19			
20						■	21			22				
■	■	■	23			24	■	■	25				■	■
26	27	28			■	29	30	31	■	32			33	34
35					36				37		■	38		
39				■	40					■	41			
42			■	43						44				
45			46		■	47			■	48				
■	■	49			50	■	■	51	52			■	■	■
53	54					55	56	■	57			58	59	60
61				■	62			63						
64				■	65					■	66			
67				■	68					■	69			

ACROSS

1. The title role in “Boris Godunov” is for a singer in this vocal range

5. This verb comes from an Old English word for “tremble”; you might do it during a temblor

10. Finish this popular saying: “It is best to be on the safe_____.”

14. The English translation for the french word: rite

15. Worrier’s worry

16. Writer Bagnold

17. Reciprocal pronoun

19. Zipped through

20. Unscramble this word: suimtm

21. The English translation for the french word: ÈnormitÈ

23. Yesterday, so to speak

25. Wished undone

26. The English translation for the french word: inane

29. The English translation for the french word: opter

32. Unscramble this word: serds

35. The English translation for the french word: rat de bibliothËque

38. To the ___ degree

39. Mysterious: Var.

40. Science

41. The English translation for the french word: stoa

42. Whiz

43. Toward that place; in that direction

45. Finish this popular saying: “The end justifies the_____.”

47. Finish this popular saying: “Waste not want_____.”

48. The last names of bestselling authors Arthur & Alex were both pronounced this way

49. This farewell word first appeared in an English text in Hemingway’s “A Farewell to Arms”

51. Zipped

53. The English translation for the french word: astÈrisque

57. Sonny boy

61. Wolfe coined the term “radical” this in a story on a party for the Black Panthers thrown by Leonard Bernstein

62. An island in southeastern New York

64. The lovely locks of a Lipizzaner

65. Uneven

66. Old Icelandic literary work

67. Finish this popular saying: “Don’t teach your Grandma to suck_____.”

68. Shut out

69. The mood was one of this as the Moon broke from its orbit & hurtled toward Earth

DOWN

1. Warner ___

2. The English translation for the french word: aÔnou

3. Watch part

4. Characteristic of or befitting a seaman

5. The English translation for the french word: quotitÈ

6. Last: Abbr.

7. Yearn (for)

8. Wail

9. Fill in the blank with this word: “Computer ___”

10. Tailoring machine

11. The English translation for the french word: incident

12. Unscramble this word: ited

13. Whirlpool

18. This Japanese term refers to second-generation Japanese-Americans, many of whom were interned during WWII

22. Fill in the blank with this word: “”How ___!””

24. A ribbed fabric used in clothing and upholstery

26. Steel girder

27. Great-___

28. Discharge bad feelings or tension through verbalization

30. Unscramble this word: oopht

31. The English translation for the french word: collant

33. Warehouse

34. Disreputable

36. Fill in the blank with this word: “”___ la la!””

37. The English translation for the french word: soude

41. Wrap in swaddling clothes

43. Rimsky-Korsakov’s “The Tale of ___ Saltan”

44. These large birds can be found in mixed herds with Guanacos in South America

46. They may appear on a tree

50. unoiled (the opposite of)

52. Buy-one-get-one-free item?

53. #1 spot

54. Uneven hairdo

55. Upset

56. Turning point?

58. Woodworking groove

59. Fill in the blank with this word: “___-European”

60. Fill in the blank with this word: “___ cheese”

63. Troop grp.

PUZZLE 37

1	2	3	■	4	5	6	7	8	■	9	10	11	12	13
14			■	15					■	16				
17			■	18					19					
20			■	21				■	22					
23			24				■	25				■	■	■
■	■	26				■	27				■	28	29	30
31	32				■	33					34			
35				■	36					■	37			
38				39					■	40				
41			■	42				■	43				■	■
■	■	■	44				■	45					46	47
48	49	50				■	51				■	52		
53						54					■	55		
56					■	57					■	58		
59					■	60					■	61		

ACROSS

1. World Match Play Championship champ a record seven times

4. Fungal spore sacs

9. Fill in the blank with this word: "___ throat"

14. Texas ___

15. Thick-coated dog

16. Yeats's work

17. Fill in the blank with this word: "___ sponte (of its own accord, at law)"

18. Witty banquet figure

20. Bygone daily MTV series, informally

21. Wild Indonesian bovine

22. Relative of a 29-Down

23. Fill in the blank with this word: "___ "Le Morte d'Arthur""

25. Tribe in Manitoba

26. Flightless bird: Var.

27. Sci. course

28. Some undergrad degs

31. Short online message

33. Bees do it: Var.

35. Yves Klein found this heavenly color a symbol of pure spirit & made works that were just a field of it

36. What fun!'

37. Crooked

38. Classic name on stage

40. Suburb of Tokyo

41. Ways around: Abbr.

42. I could ___ horse!'

43. Bottom-of-letter abbr.

44. Seuss's "Horton Hears ___"

45. Start a voyage

48. Spanish hill

51. Start of a 1957 hit song

52. Magician's name ending

53. A pharaoh vis-

55. Computer units: Abbr.

56. The English translation for the french word: niche

57. Nothing runs like a ___' (ad slogan)

58. Self: Prefix

59. Whoopi's role in "The Color Purple"

60. Right turn ___

61. Vietnam War-era org.

DOWN

1. Kodak founder

2. Mrs. Bush

3. Predawn period

4. Semitic fertility goddess

5. The Earl of Sandwich, for one

6. Slangy greetings

7. Guy Lombardo's "___ Lonely Trail"

8. Pioneering anti-AIDS drug

9. Oh, please, that's enough'

10. We love ___ you smile' (old McDonald's slogan)

11. Softens in water, in a way

12. Go-aheads

13. Nickname for JosÈ

19. Unscramble this word: elrmey

24. Virginia's ___ River

25. Let's wax philosophical & wonder why Rousseau left only fragments of his opera about Daphnis & her

27. King ___ (dangerous snake)

28. Three-stringed instruments

29. Warehouse

30. Print tint

31. Trails off

32. About 40 degrees, for N.Y.C.

33. Unscramble this word: oopht

34. Workers with dogs, maybe

36. Warren Buffett, for one

39. English churchyard sight

40. Like neglected muscles

43. Winnie-the-Pooh's donkey friend

44. Japanese beer brand

45. Comparatively gritty

46. Preparing to bloom

47. Rolls

48. Pres., to the military

49. Combined, in Compi

50. O.T. book

51. Call in the game Battleship

54. Without further ___

PUZZLE 38

ACROSS

1. Standing by

6. Stench

10. Nickname of Israel's Netanyahu

14. Where le pr

15. Dweeb

16. Step ___!'

17. Kipling short story, with "The"

19. Chairmen often call them: Abbr.

20. Some nest eggs

21. White House monogram

22. Steakhouse order

24. Fill in the blank with this word: "___ possidetis (as you possess, at law)"

25. To the ___ power

26. The Lizard constellation

27. Spacecraft segment

29. Sinclair Lewis's ___ Timberlane

30. Fill in the blank with this word: "___ in Charlie"

31. Siesta for a feline?

33. Popular theater name

34. Home of Michigan State

38. Zealous

39. Toodles!'

40. Celtic sea god

41. Was ___ hard on them?'

43. Virgin Island that's 60% national park

47. Kind of rice used in risotto

49. Heave-___ (dismissals)

50. Wharton grad's aspiration, maybe

51. The English translation for the french word: intelligence

52. Big inits. in paperback publishing

53. The English translation for the french word: troupeau

54. City on the Rhein

55. See 22-Across

58. Fill in the blank with this word: "Comic strip "___ & Janis""

59. The Naturist Society sponsors this type of recreation week, a chance for you to let it all hang out

60. Blakley of 'Nashville'

61. Love ___

62. The ruler of Qatar is known by this 4-letter title

63. I bid you ___ farewell'

DOWN

1. The English translation for the french word: osmium

2. On the verge of

3. Like the Taj Mahal's marble

4. Those Animals Frighten Me! : Ailurophobia

5. Saison d'___

6. General ___ chicken

7. The English translation for the french word: cintre

8. Domingo, for one

9. Hindu precepts

10. Ice cream dessert

11. Office squawker

12. 1975 Joni Mitchell hit

13. "Piece of cake!"

18. "Fiddler on the Roof" setting

23. Suffix with robot

25. About 40 degrees, for N.Y.C.

26. The English translation for the french word: retomber

28. School in La Jolla: Abbr.

29. Echo producer

32. Queen Amidala's home in "Star Wars" films

33. Actress Gilbert of "Roseanne"

34. From here to eternity

35. Hoops bloopers

36. Step on it

37. Have-___ (poor people)

38. Fill in the blank with this word: "___ king crab"

41. Fill in the blank with this word: ""Just Another Girl on the ___" (1993 drama)"

42. Herbal tea

44. Setting of muchas islas

45. Men of the haus

46. Signed an agreement?

48. With 46-Down, words finishing "Ready ___, here ___"

49. June of 'Scudda Hoo! Scudda Hay!'

52. Knots

53. Fill in the blank with this word: "Dog : paw :: horse : ___"

56. The English translation for the french word: vagabond

57. Training ___

PUZZLE 39

ACROSS

1. Where Moses got the Ten Commandments

8. USA alternative

11. What's funded by FICA, for short

14. Fill in the blank with this word: “English poet Coventry ___, who wrote “The Angel in the House””

15. Fill in the blank with this word: “___ in apple”

16. Stamp not reqd.

17. Surf serving #4

19. Source of some rings

20. Per ___ (daily)

21. World production of this is now 84 million barrels a day, up about 10 million barrels from 1997

22. Wood: Prefix

23. 2000 site

27. Way from Syracuse, N.Y., to Harrisburg, Pa.

28. 1965 #1 hit by the Byrds

29. Fill in the blank with this word: “”If ___ Would Leave You””

30. Word on a dipstick

31. Fill in the blank with this word: “Architect ___ Ming Pei”

33. Some hip-hop women

34. the hole in a woodwind that is closed and opened with the thumb

36. Phylicia of stage and screen

39. Chopin’s “Butterfly” or “Winter Wind”

40. The English translation for the french word: viscÈral

43. Work translated by Pope

44. Windsor’s prov.

45. Natal native

46. Unwelcome sight in the mail

50. Henley Regatta setting

51. Groove-billed ___

52. Fill in the blanks with these two words: “”Whatcha ___?””

53. Sir ___ McKellen (Gandalf portrayer)

54. Spanish sherry

58. Yabba dabba ___!’

59. What’s missing from a KO?

60. Patella

61. Most miserable hour that ___ time saw’: Lady Capulet

62. Reply facilitator: Abbr.

63. Snags

DOWN

1. UK legislators

2. Tit for ___

3. Star-___

4. Nothing’s broken!’

5. Phrase in some apartment ads

6. Fill in the blank with this word: “”You ___ Lucky Star” (1935 #1 hit)”

7. Weapon for Iraqi insurgents: Abbr.

8. One who knows “the way”

9. Call that may complete a full count

10. Wind dir.

11. You may want to stop reading when you see this

12. Member of a 1990s pop quintet

13. Shelley poem

18. Home of Notre Dame

22. Fill in the blank with this word: “”C’est ___””

23. Veracruz Mrs.

24. Wall Street earnings abbr.

25. Mr. ___, radioactive enemy of Captain Marvel

26. Fill in the blank with this word: “En ___ (by the rules: Fr.)”

31. Fill in the blank with this word: “”If ___ nickel...””

32. Fix a squeak

33. Fill in the blank with this word: “”What a ___!” (beach comment)”

34. Rodgers and Hart’s “___ Love”

35. Tutsi foe

36. Swimmer’s fear

37. Song written by Queen Liliuokalani

38. Yes sir!,’ south of the border

40. Wild llama

41. Where Einstein was born

42. Ram, in Ramsgate

44. Wife of Paris, in myth

45. They’re great on Triple Letter Scores

47. This type of radiation, & letter of the alphabet, hulked out Bruce Banner

48. Patsy’s pal on “Absolutely Fabulous”

49. Undersides

54. The Unsers of Indy

55. Damascus’s land: Abbr.

56. Peace Nobelist Kim ___ Jung

57. Special ___

PUZZLE 40

ACROSS

1. Fill in the blank with this word: “Burkina ___”

5. Henry ___ Lodge

10. Formal hat, informally

14. Strike ___ blow

15. Fill in the blank with this word: “Clifford ___, “Awake and Sing!” dramatist”

16. Large hall

17. Remake about a holy person’s slip?

20. They go into overtime

21. What’s happening

22. Step ___!’

23. Some Iroquois

25. Easy-to-prepare, as cheesecake

28. Phlebitis targets

29. Lover: Suffix

30. Take ___ of (sample)

31. Purveyor of nonstick cookware

35. Fill in the blank with this word: “”___ help a lot!””

36. She betrayed Samson

39. QuÈbec's ___ d'OrlÈans

40. Vetoes

42. Trails

43. Tied, as a French score

45. Surprise Symphony' composer

47. They may take a few yrs. to mature

48. The English translation for the french word: point díaccËs

51. Copter's forerunner

52. Fill in the blank with this word: ""Are you ___?""

53. Summer chirpers

57. Line from a Copland "Portrait"

60. Seed covering

61. Martinique volcano with a violent 1902 eruption

62. Tucson-to-New Orleans route

63. Series of legis. meetings

64. Waldorf ___

65. Kit ___ (Hershey bars)

DOWN

1. Yom Kippur ritual

2. Jai ___

3. Peau de ___ (soft fabric)

4. Soccer goof

5. a socialist who advocates communism

6. On ___ (trying to lose)

7. Porgy and ___'

8. Where to play favorites?: Abbr.

9. Taoism founder Lao-___

10. Russian reactionary

11. Unlike dirt roads

12. "Nadja" actress L

13. Shoemakers' leather strips

18. Failed completely

19. Coordinate in the game battleships

23. Fill in the blank with this word: "___-de-boeuf (oval windows)"

24. Thatching palm

25. It's in the back row, right of center

26. Fill in the blank with this word: "" ___ be in England...""

27. When you wait for something, you do this to your time

28. Winston Churchill flashed it

30. Luck Be ___ Tonight'

32. Order

33. Pas ___ (gentle ballet step)

34. Wranglers alternative

37. Actor Baskin of "Air Force One"

38. Son of William the Conqueror

41. informal terms for money

44. Bum

46. Wowed

47. Like the Marquis de Sade or the Duke of Earl

48. Dos

49. Fill in the blank with this word: ""Bird on ___," 1990 Mel Gibson movie"

50. Smiley of PBS

51. Helmetlike flower petal

53. Fill in the blank with this word: ""A View to a ___" (Bond film)"

54. Whit

55. Unscramble this word: ited

56. Workers need them: Abbr.

58. Velocity meas.

59. Fill in the blank with this word: "___ Today (teachers' monthly)"

PUZZLE 41

ACROSS

1. Winter weather, in Edinburgh

4. What's spread on a spreadsheet

8. Contemptible sneaks

13. One of a series of joint Soviet/U.S. space satellites

14. Small islands

15. Via ___ (Roman road)

16. Pen pals?

18. Twit

19. Fill in the blank with this word: “”___ Warning” (“Das Rheingold” aria)”

20. U.S.M.C. noncoms

21. Width measure

22. Fill in the blank with this word: “ì___ boom bah!î”

25. Winning sch. in the second Sugar Bowl

26. Shelley's “___ Skylark”

28. Working best, as a trick

30. Years on end

31. World Match Play Championship champ a record seven times

32. Modern travel aids

34. Winnebagos, for short

35. The English translation for the french word: soma

38. Weight allowances

39. Warhol's "___ of Six Self-Portraits"

40. Santa ___, Calif.

41. Without dissent

43. Works for an ed.

44. My ___, Vietnam

45. Stews

49. Suffix with robot

50. Summer, in France

51. Fill in the blank with this word: "___ and snee"

52. Young fellow

53. Hamilton' actress ___ Elise Goldsberry

55. "Soap" family

57. Made off with

58. Multicapable

62. Watch again

63. Fill in the blank with this word: """___ Anything" ("Oliver!" song)"

64. Viennese-born composer ___ von Reznicek

65. Fill in the blank with this word: "Edgar Bergen's Mortimer ___"

66. Willie of the 1950s-'60s Giants

67. Sponge (up)

DOWN

1. No ___!'

2. Got me'

3. Vase's handle

4. Big jet

5. Popular cable channel

6. Norse war god

7. Smart ___ (wise guy)

8. Some bygone roadsters

9. Take down ___ (humble)

10. Paints like Pollock

11. Liquefy

12. Reply facilitator: Abbr.

13. Rounded end

17. Native of the central Caucasus

20. Tartan wearer

23. Question to a consumer watchdog

24. Popular tropical destination

25. Wobbles on the edge

27. Fill in the blank with this word: "___ prof."

29. The Missing Drink : High ____ rose

33. Fill in the blank with this word: """The Longest Day" director ___ Annakin"

35. Reindeer-herding people

36. In the movies

37. She kneads people

39. So that's ___?'

41. Took in, perhaps

42. Unsettled feeling

46. Last month

47. Nickname for Leo Durocher

48. Cut

54. What an A is not

56. Suffixes with sultan

57. Yearbook sect.

58. When paired with vigor, it signifies exuberance

59. Soprano Christiane ___-Pierre

60. Fill in the blank with this word: "At a low ___"

61. Fill in the blank with this word: "___ pad"

PUZZLE 42

ACROSS

1. Theme of this puzzle

5. Where the Althing sits: Abbr.

9. Wanting no more

14. Fill in the blank with this word: “”... gimble in the ___”: Carroll”

15. go (the opposite of)

16. Significant ___

17. This ___ outrage!’

18. The English translation for the french word: suie

19. Serviceable

20. 1965 #1 hit by the Byrds

23. Some flawed mdse.

24. Fill in the blank with this word: “”___ was saying Ö”””

25. Wildlife threat, briefly

28. “Ah, Wilderness!” mother

30. Bony

32. Pol. designation for Gov. Jeanne Shaheen

33. Moon material, supposedly

37. High-tech transmission

39. Three-way joint

40. Transcript stats

41. 13

46. Tiny energy units, for short

47. Fill in the blank with this word: "___ del Fuego"

48. Fill in the blank with this word: "___ double life"

50. Series

51. In ___ of

53. Old-fashioned letter opener

54. Call after a hit by 17-Across

59. Singer with a falsetto

62. Fill in the blank with this word: "___ Lemaris, early love of Superman"

63. Poet Mandelstam

64. Venusian, e.g.

65. Radio host Don

66. The Mounties: Abbr.

67. Put a new price on

68. Part of S.E.C.: Abbr.

69. Two out of twenty?

DOWN

1. Twerp

2. Pitts of silent film

3. Letter-shaped construction piece

4. With 47-Down, title for this puzzle

5. John Hancock, e.g.

6. The Rockies' ___ Field

7. Designer von Furstenberg

8. "Little" girl of old comics

9. Acidulous

10. Memo abbr.

11. Summer comfort stat

12. Fill in the blank with this word: "Electric ___"

13. Rap's Dr. ___

21. Fill in the blank with this word: "___ function"

22. Summer Games org.

25. Espied Godiva, e.g.

26. any of several plants of the genus Manihot having fleshy roots yielding a nutritious starch

27. Fill in the blank with this word: ""Lord, ___ this food" (grace words)"

28. As a matter of fact: Fr.

29. Millennial Church member

31. two-year-old sheep

32. Things short people have?

34. You may be asked to arrive 90 mins. prior to this

35. Want ad inits.

36. Pince-___

38. Generation ___ (twenty-something)

42. What a slasher usually gets

43. Wood cutters

44. Undemocratic tendency

45. Tiber tributary whose name means "black"

49. Middle of the riddle

52. Ancient Greek tongue: Var.

53. The English translation for the french word: pincer

54. Intestinal parts

55. L'___ Vogue, Italian fashion magazine

56. Verb ending

57. Fill in the blank with this word: ""The ___ of the Ancient Mariner""

58. Omar of "The Mod Squad," 1999

59. Not the usual spelling: Abbr.

60. Pale ___

61. Kiddie ___

PUZZLE 43

ACROSS

1. Stew container

6. Terhune's "___ Dog"

10. What's missing from a KO?

13. Fill in the blank with this word: “”___ bleu!””

14. Newswoman Magnus et al.

15. Width measure

16. Ride for 007

18. N.B.A. on ___ (sports staple since 1989)

19. When you pick up a tab, you do this to "the bill"

20. Strains

22. Wyo. hours

24. Relative of "i.e."

26. Crocus or freesia, e.g.

27. Two qtrs.

28. Not thinking well

30. .___

32. Worked on the Street

34. Evangelist

37. To go, to Godot

38. Lao-___

39. Pancreatic hormone

40. Traditional site of Jesus' crucifixion

42. "Swish!"

43. Fill in the blank with this word: "Dutch painter Gerard ___ Borch"

44. Wild cards, in a certain game

46. The English translation for the french word: cÈ

47. Free ___ (total control)

49. Fill in the blank with this word: ""So ___ me!""

50. Words to masseuses

51. Fixes a dress

53. Teri ___, Best Supporting Actress nominee for 'Tootsie'

55. Put ___ good word for

56. Stephen King novel

62. Third-century year

63. To be, in Toledo

64. Fill in the blank with this word: "___-one (long odds)"

65. Fill in the blank with this word: "___ Monroe, "Green Acres" role"

66. What's ___ you?'

67. Turned

DOWN

1. Union-busting grp.?

2. Fill in the blank with this word: "___ Tafari (Haile Selassie)"

3. Natl. Boss Day, ___ 16

4. Video game heroine Lara ___

5. Numbers game

6. Pres., e.g.

7. Small island

8. Mercury and Saturn

9. Richards : Moore :: Grant : ___

10. Informal invitation

11. Word on a wall, in the Bible

12. Trawling equipment

14. Stay home for supper

17. Tearful

21. Most pleasing

22. Wine sometimes blended with Cabernet Sauvignon

23. Pie cutter's oxymoron

25. Irregular courses

27. Without women

28. Prefix in hematology

29. Skip to My ___'

31. Set-tos

33. Temperature unit

35. There have been 12 popes with this devout name, the first during the second century, the last from 1939 to 1958

36. The English translation for the french word: rite

38. Summer comfort stat

41. Stretched to the limit

42. Substitute players

45. More monumental

48. Rock and Roll Hall of Fame designer

50. Soviet co-op

51. Fill in the blank with this word: "Costa ___"

52. Bottom-of-letter abbr.

54. Word of disappointment

57. 1965 #1 hit by the Byrds

58. Fill in the blank with this word: "___ TomÈ"

59. Australian singer Christine

60. Fill in the blank with this word: "___ 1 (Me.-to-Fla. highway)"

61. Tenth letter of the Hebrew alphabet

PUZZLE 44

ACROSS

1. Wide-headed fastener

5. Very

9. Solution strength: Abbr.

13. Res ___ loquitur (legal phrase)

14. Rep. ___ Hastings of the House intelligence committee

15. They're often bitter

16. Irish playwright who wrote "Cock-a-Doodle Dandy"

18. Former N.F.L. great Junior ___

19. Economic stat.

20. The English translation for the french word: utÈrus

21. E-flat, on a Steinway

23. Victorian taxi

25. Restrain

26. Actor Hickman showed boredom

32. L-___ (treatment for parkinsonism)

35. Fill in the blank with this word: ""Fur ___" (Beethoven dedication)"

36. Grand ___ Opry

37. Short race, for short

38. The English translation for the french word: DAS

39. Three-time N.H.L. All-Star Kovalchuk

40. Lew Wallace's "Ben-___"

41. The English translation for the french word: haÔku

43. Writer's Market abbr.

44. Entrance requirement, maybe

48. Fill in the blank with this word: ""Let ___!""

49. The English translation for the french word: couvreur

53. The Preserver, in Hinduism

56. Set ___ (embark)

58. One in the charge of un instituteur

59. Thought: Prefix

60. *Creation made with a bucket and shovel

63. Wall-plastering material

64. Fill in the blank with this word: "___-high boots"

65. Secular

66. You and who ___?' (fighting words)

67. View in northern Italy

68. African fox

DOWN

1. New York University's ___ School of the Arts

2. Salt-___ (rap trio)

3. Track star Bolt

4. Fill in the blank with this word: "Farmer's ___"

5. Verbal assault

6. Campus 100 miles NW of L.A.

7. The English translation for the french word: cÈ

8. Primes

9. Tropical nut

10. Olive genus

11. Fill in the blank with this word: "___ East"

12. Burger topper

14. In ___ (unconscious)

17. It hurts!'

22. Union member

24. Wyo. neighbor

25. Troubled capital

27. Possible 'Got milk?' reply

28. Body of water in a volcanic crater, for one

29. Vincent Lopez's theme song

30. Golf innovator Callaway and bridge maven Culbertson

31. "My stars!"

32. Qatar's capital

33. Move ___' (Curtis Mayfield song)

34. Vivacious

39. Lenin's "What ___ Be Done?"

41. Coordinate in the game battleships

42. Early capital of Georgia

45. Words served with Honey?

46. Psychoanalyst Fromm

47. Fa followers

50. Brine-cured cheeses

51. Fill in the blank with this word: "___ Island National Monument"

52. Fill in the blank with this word: "1920s-'60s Tennessee congressman B. Carroll ___"

53. The English translation for the french word: vil

54. TV's "American ___"

55. Wet septet

56. Use tiny scissors

57. Lead ___ life

61. Sports org. with the Calder Cup

62. Radical 1970s grp.

PUZZLE 45

ACROSS

1. Fill in the blank with this word: "___ ligation"

6. Convocation of witches

11. Some pops

14. Standing by

15. Turf

16. Want-ad letters

17. a pitched battle in which American revolutionary troops captured Fort Ticonderoga from the British in 1775

19. Where the Paran

20. Without restraint

21. Fill in the blank with this word: "___ Tamid (synagogue lamp)"

22. Band with the 1999 hit "Summer Girls"

23. Loosen, in a way

26. a gaping grimace

28. There's ___ in team'

29. Ranger, e.g.

33. Money machine mfr.

34. Wore away

35. Chopin's "Butterfly" or "Winter Wind"

36. Get ___ reception

39. Fill in the blank with this word: "___ the way"

41. Fill in the blank with this word: "Everything ___ place"

43. French antiseptic

44. Up on deck

46. Crooked

47. The English translation for the french word: simbleau

48. Void, in Vichy

49. Replace a wooden pin

51. Taina who was one of Les Girls, 1957

52. Lawn game

55. Just about

57. Fill in the blank with this word: "1960s Elvis-style singer ___ Donner"

58. Fill in the blank with this word: "___-la-la"

60. Fill in the blank with this word: "___-free"

61. Fill in the blank with this word: ""Lord, is ___?": Matthew"

62. Utter gibberish

67. The 21st, e.g.: Abbr.

68. Winemaking village east of Verona

69. Twist

70. Yacht's dir.

71. Windows predecessor

72. Sell online

DOWN

1. Youngster

2. Verse starter?

3. Wite-Out manufacturer

4. Flip ___ (decide by chance)

5. See 53-Down

6. Some shoes ... and a feature of this puzzle's theme

7. Plaza abbr.

8. Fill in the blank with this word: """___ giorno!"""

9. Tool with a cross handle

10. Sovereign of yore

11. One unlikely to punk out

12. Where Prince Philip was born

13. Western lilies

18. Attends to a detail

23. Not suitable

24. ___ a Stranger' (Olivia de Havilland film)

25. Eastern European hill?

27. Year in Diocletian's reign

30. The English translation for the french word: lunaire

31. Roman title

32. Surface again, as a road

37. Newbery-winning author Scott ___

38. Long-limbed, as a model

40. The "E" of N.E.A.: Abbr.

42. Fill in the blank with this word: "2001 Economics Nobelist Michael ___"

45. Undemocratic tendency

50. With 50-Down, speaker of the quotation

52. Fill in the blank with this word: "Fanny ___ of the Ziegfeld Follies"

53. Writer Joyce Carol ___

54. Composer Dohn

56. Pitch ___ (prepare to camp)

59. Nav. ___

60. Tom Jones's "___ a Lady"

63. Omne vivum ex ___ (all life [is] from eggs: Lat.)

64. Sports org.

65. Fill in the blank with this word: "Dernier ___"

66. Wright wing?

PUZZLE 46

ACROSS

1. Yom Kippur service leader

6. Repeated interjection in the Rolling Stones' "Miss You"

10. Union foe

14. Fill in the blank with this word: "___ the hole"

15. Fill in the blank with this word: “”___-daisy!””

16. Roman statesman ___ the Elder

17. PBS policy

18. Sub

19. Fill in the blank with this word: "Astronomy's ___ cloud"

20. Classic Miles Davis album ... or a hint to the start of 17-, 22-, 37- or 45-Across

22. They often precede la's

23. *Group with the 2000 #1 hit "It's Gonna Be Me"

24. Tiny time unit: Abbr.

26. Laid bets at a casino

29. Cello feature

33. Part of USPS: Abbr.

37. Whack

38. Detailed, old-style

39. Fill in the blank with this word: "___ Cologne (skunk of old cartoons)"

40. Abbr. after Ted Kennedy's name

42. Unscramble this word: gang

43. Some October babies

45. Poker supplies

46. With 8-Down, source of an ethical dilemma

47. Wordless song: Abbr.

48. Under-the-sink fitting

50. Singer Brickell

52. Fill in the blank with this word: "___ artery"

56. Vestments, e.g.

59. Republication

63. Tabby talk

64. Fill in the blank with this word: ""The House Without ___" (first Charlie Chan mystery)"

65. Vulcan portrayer

66. Trollope's "Lady ___"

67. Fill in the blank with this word: "___ East"

68. Need for the winner of a Wimbledon men's match

69. Union member

70. White-tailed eagle

71. Successful job applicant

DOWN

1. What stripes may indicate

2. Flip ___ (decide by chance)

3. You don't know ___'

4. Fussbudget

5. One way to break out

6. Bushman's home

7. Volkswagen competitor

8. Royal fern

9. Pumps up

10. Sausage-wrapped British breakfast dish

11. Openness

12. Gillette ___ Plus

13. Nonhuman co-hosts of TV's "Mystery Science Theater 3000"

21. King and queen

25. Red ___ (young amphibian)

27. Open a New Window' musical

28. Difference in days between the lunar and solar year

30. Sheik ___ Abdel Rahman

31. Watercolorist ___ Liu

32. Wired

33. Songs for one

34. Fill in the blank with this word: "___ chief (publ. honcho)"

35. You've got 24 of these in the front of your chest, protecting your inner organs

36. Facet joints connect them

38. See 34-Across

41. White Sulphur ___, W. Va.: Abbr.

44. Suffix with tank

48. Cheater, perhaps

49. Statue base

51. Witless

53. You're looking at him!'

54. Archer, at times

55. Robert of Broadway's "My Fair Lady"

56. Transcript stats

57. West End classic "Charley's ___"

58. Cell stuff that fabricates protein, for short

60. Withdraw gradually

61. Fill in the blank with this word: "BrontÎ's "Jane ___""

62. Where Loews is "L"

PUZZLE 47

ACROSS

1. Unscramble this word: gang

5. Tarantula-eating animal

10. Vitamin a.k.a. riboflavin

14. Latin hymn "Dies ___"

15. The English translation for the french word: mimique

16. Wilt thou not chase the white whale?' speaker

17. Unleashes

19. Bring (up) from the past

20. Right back ___!'

21. Worse in quality, slangily

23. Linda Ellerbee's "___ It Goes"

26. Fill in the blank with this word: "Day ___"

27. TV announcer Hall

30. Tournament sit-out

31. Waiting in the wings

35. Wreath

36. Northwest Terr. native

38. Eightfold

39. Cashier's error, as suggested by 17-,

22-, 47- and 58-Across?

43. Much Mongolian geography

44. Fill in the blank with this word: "___ sponte (of its own accord, at law)"

45. Fill in the blank with this word: ""Am ___ risk?""

46. Kind of a drag

47. Went off

49. The English translation for the french word: cÈ

50. Like ___

52. Sufficiently old

54. Pop star who was a 1990's Mouseketeer

58. Weapons check, in brief

62. Rider-propelled vehicle, for short

63. Least desirable wharves?

66. The Bible's Garden of ___

67. Fill in the blank with this word: ""___ Cassio!": Othello"

68. Villainous resident of Crab Key island

69. Sugar suffixes

70. Canio's wife in "Pagliacci"

71. Works of Homer

DOWN

1. Bhagavad-___

2. Stuck in ___

3. Fill in the blank with this word: "___ a soul"

4. Salami choice

5. Year that Eric the Red was born, traditionally

6. Source of some rings

7. Shell alternative

8. Mrs. Addams, to Gomez

9. They're served with spoon-straws

10. 2003 movie involving Christmas Eve robberies

11. Fill in the blank with this word: ""___ the brinded cat hath mew'd": "Macbeth""

12. Living ___ (what an employer is asked to pay)

13. Upper: Ger.

18. Beach application

22. Time

24. Textile workers

25. Zaire's Mobutu ___ Seko

27. Tool for some group mailings

28. Fill in the blank with this word: "Al ___ (not too soft)"

29. Like Waldorf salad apples

32. Viking garment

33. These primitive plants are usually grouped according to color: brown, red, blue-green, golden...

34. Quarterback Rodney

37. Basketball coach Jones and others

38. Florida's ___ National Forest

40. They're just what you think

41. The second "R" in J. R. R. Tolkien

42. The English translation for the french word: turf

48. Fill in the blank with this word: "___ even keel: 2 wds."

51. Victors' cry

52. Rowed

53. The English translation for the french word: Ètude

54. Onetime Chevy subcompact

55. Sheepskin alternatives, for short

56. Victor Nu

57. Went on

59. Tombstone name

60. Fill in the blank with this word: "___-Tibetan languages"

61. General ___ chicken (Chinese menu item)

64. Woeful

65. Fill in the blank with this word: "___-la-la"

PUZZLE 48

ACROSS

1. “Through the Looking-Glass” antagonist

5. Understands

9. Lawyers: Abbr.

13. City of northern Finland

14. Fill in the blank with this word: “Crazy as ___”

16. Letters on a R

17. Veteran journalist ___ Abel

18. Fill in the blank with this word: “___-Grain cereal bars”

19. The last Mrs. Chaplin

20. Annoy people by blowing a fan on them?

23. Use a surgical beam

24. Theroux’s “The Happy ___ of Oceania”

25. Roman square?

28. Army helicopter

32. Concerned query

34. Motel freebie

35. Widen, in a way

39. No-good, awful frozen waffle?

43. Fill in the blank with this word: “”Join the ___ ””

44. Where the outboard motor goes

45. Tidal flood

46. Like Seattle’s skies often

49. Fill in the blank with this word: “China’s Sun ___-sen”

50. 1957 #1 song

54. Novice: Var.

56. Stop order?

63. Fill in the blank with this word: “”Yo, ___!””

64. West Indies native

65. Diggs of ‘Rent’

66. U.K. military medals

67. Spotted ___

68. Yield, as interest

69. Zip

70. River isles

71. Fill in the blank with this word: “___ place”

DOWN

1. Film director Nicolas

2. Fill in the blank with this word: “Faulkner’s femme fatale ___ Varner”

3. Mae West role

4. Suppress

5. Send far, far away, maybe

6. Sitting muscles

7. The lady ___ protest too much’: Shak

8. Maelstrom

9. Swears

10. Plowed layer

11. Unit for a lorry

12. W.W. I French fighter planes

15. Wellness org.

21. The English translation for the french word: fou

22. Apt. feature, in the classifieds

25. Ball V.I.P.’s

26. Work hard

27. Wheat variety

29. Meshlike

30. Natl. Boss Day, ___ 16

31. Fill in the blank with this word: “Confit d’___ (potted goose)”

33. Where to play favorites?: Abbr.

35. Old New Yorker cartoonist Gardner ___

36. Rich, as food

37. You’ll want to munch on petha & gazak, signature sweets of this Taj Mahal city

38. Bubbly name

40. Soldier

41. U.S. ___

42. This was Indira Gandhi’s maiden name (her father was India’s first P.M.)

46. What a floozy might show off

47. Fill in the blank with this word: “”If ___ broke...””

48. Fill in the blank with this word: “”The Lion King II: ___ Pride””

50. Huey, Dewey, Louie, Donald and Daisy

51. 2008 documentary about the national debt

52. Some choice words

53. Weapons check, in brief

55. Prefix with sclerosis

57. Scot’s exclamation

58. Fill in the blank with this word: “___ con Dios (Spanish farewell)”

59. Riley’s “___ Went Mad”

60. The Godfather’ co-star

61. This instrument of the cult of Apollo lent its name to the type of poetry it accompanied

62. Twilights, poetically

PUZZLE 49

ACROSS

1. Fill in the blank with this word: "___ congestion"

6. On ___ (exulting)

11. Mil. pilot's award

14. Yo, she was Adrian

15. Threepeater's threepeat

16. Sports org. for nonprofessionals

17. Popular fictional 31-Down

19. To ___ is human ...'

20. Stands for viewings

21. Required

23. Peak on the eastern edge of Yosemite Natl. Park

26. Liqueur flavoring

27. Unscramble this word: htglen

28. Quite the success

29. Untilled tract

30. Fill in the blank with this word: "___-de-boeuf (oval windows)"

32. With 64-Across, 1-/66-Across movie of 2005

35. French river or department

37. Cool, very red celestial body

39. Nagy of Hungary

40. While away, as time

42. Father and victim of Oedipus

44. Whitman's "A Backward Glance ___ Travel'd Roads"

45. Oklahoma athlete

47. Unscramble this word: rsecte

49. Make sure

51. Reveals one's feelings: 2 wds.

52. Fill in the blank with this word: "___ Park, historic home near Philadelphia"

53. Upholstery fabric

55. Ticket abbr.

56. Influential group

61. Shamus

62. Weapons check, in brief

63. Big tournaments for university teams, informally

64. Old letter

65. Rest room sign

66. You may use a stick for these

DOWN

1. To the ___ power

2. Fill in the blank with this word: ""___ approved" (motel sign)"

3. Type of 35mm camera

4. Source of inflation after a crash

5. Let have it

6. Suffix with origin

7. Website statistic

8. TV's Cousin ___

9. Musician with the first record formally certified as a million-seller

10. Words before "signed, sealed, delivered" in a Stevie Wonder hit

11. Revived

12. Fill in the blank with this word: ""___ it from me...""

13. The English translation for the french word: guÈri

18. The English translation for the french word: paonne

22. Writing by Montaigne

23. Fr. girls

24. Start a hole

25. Human Genome Project, e.g.

26. Fill in the blank with this word: "Allegro ___ (very fast)"

28. ___ boy

31. Fill in the blank with this word: "___ of Langerhans"

33. The English translation for the french word: ramper

34. What's the frequency, Kenneth?--& make sure it's in this unit equal to one cycle per second

36. Short online message

38. Snuffy Smith, for one

41. Sending to one's fate

43. One side of traffic

46. Dolce far __

48. The English translation for the french word: imposer

49. 1994 sci-fi epic

50. Went wrong

53. Fill in the blank with this word: "Eye ___"

54. Waste of a meal

57. Pitcher Robb ___

58. Mike Ovitz's former co.

59. Fill in the blank with this word: ""Knots Landing" actress ___ Park Lincoln"

60. Winding road shape

PUZZLE 50

1	2	3	4	5	■	6	7	8	9	■	10	11	12	13
14					■	15				■	16			
17					■	18				■	19			
■	■	■	20		21					22				
23	24	25	■	26			■	■	27					
28			29	■	30		31	32			■	33		
34				35	■	■	36				37	■	■	■
■	■	38			39	40						41	■	■
■	■	■	42					■	■	43			44	45
46	47	48	■	49				50	51	■	52			
53			54			■	■	55		56	■	57		
58						59	60				61	■	■	■
62				■	63				■	64		65	66	67
68				■	69				■	70				
71				■	72				■	73				

ACROSS

1. Louis of the F.B.I.

6. Unscramble this word: ciph

10. Swear to

14. Fill in the blank with this word: "2010 Olympic ice dancing gold medalist ___ Virtue"

15. Fill in the blank with this word: "___ brace (device used to immobilize the head and neck)"

16. Formal discourse

17. Strained

18. Unscramble this word: liar

19. Prefix with angular

20. 50's TV catch phrase transmuted

23. You reap what you ___'

26. Stay-at-home ___

27. Lug: Var.

28. Swift Malay boat

30. New York Senator

33. Fill in the blank with this word: "Fiddle-de-___"

34. Waterwheel

36. Movie critic Roger

38. With 86-, 17- and 91-Down, how to "illustrate" this puzzle

42. The English translation for the french word: haÔku

43. Online shopping center

46. What a patrol car might get, for short

49. Photographer's support staff

52. Witches' ___

53. Start of titles by Auden, Wharton and Paine

55. The Orient Express, e.g.: Abbr.

57. Venomous snake

58. Ones who can handle adversity

62. Ancient city with remains near Aleppo

63. Mussorgsky's 'Pictures ___ Exhibition'

64. Tiler's need

68. Gulf of ___, body of water next to Viet Nam

69. Zaire's Mobutu ___ Seko

70. Dwarf warrior in 'The Lord of the Rings'

71. Chevrolet model

72. Steely Dan's "___ Lied"

73. U.S.M.C. noncoms

DOWN

1. Way to send docs. electronically

2. Theologian's subj.

3. Fill in the blank with this word: ""Humanum ___ errare""

4. Tallinn native

5. Muslim judge

6. chromatic purity: freedom from dilution with white and hence vivid in hue

7. Fill in the blank with this word: "Den ___, Nederland"

8. Of the hipbone: Prefix

9. Person with opinions

10. Like some chambers

11. Watched

12. Actual title of the 1979 #1 hit known as "The Pi

13. Make more presentable, as a letter

21. Tenth letter of the Hebrew alphabet

22. Long-shot candidate

23. "Wheel of Fortune" option

24. Siglo de ___ (epoch of Cervantes)

25. With 33-Across, anagrams and puns (or parts hidden in 17-, 24-, 44- and 51-Across)

29. Seal's opening?

31. Got together

32. Fill in the blank with this word: ""The ___ Daba Honeymoon""

35. Strip's cry of disgust

37. Fill in the blank with this word: "___ of the Unknowns"

39. Tun

40. Word with bum or bunny

41. The English translation for the french word: Nara

44. Fill in the blank with this word: "___ Cayes, Haiti"

45. It

46. The English translation for the french word: tÈmoigner

47. The English translation for the french word: phobie

48. 1950's sitcom starring Ethel Waters

50. Fill in the blank with this word: "___ Islands (Scapa Flow locale)"

51. What makes MADD mad

54. Fill in the blank with this word: ""It's only ___!""

56. Stethoscope users, at times

59. Fill in the blank with this word: "Cup ___ (hot drink, informally)"

60. Where hops are dried

61. Hindu titles

65. Texter's 'Wow!'

66. Last: Abbr.

67. Word repeated in Emily Dickinson's "___ so much joy! ___ so much joy!"

Solutions

Puzzle Solution 1

H	E	M	E		S	I	T	U				D	E	I
E	M	M	E		O	N	E	G	A		D	E	E	R
B	I	L	L	H	U	D	S	O	N		E	L	E	V
E	R	I		S	T	E	T		T	N	O	T	E	S
			K	N	E	X		R	A	I	D			
S	E	A	A	I	R		P	E	R	S	O	N	A	E
O	M	A	R			T	O	B	E		R	I	S	K
N	O	L	A		I	O	T	A	S		I	S	P	Y
I	N	T	O		R	H	O	S			Z	A	C	H
C	O	O	K	B	O	O	K		K	E	E	N	A	N
			E	I	N	E		K	A	N	S			
L	A	M	B	D	A		D	I	V	S		A	R	M
A	B	A	A		G	E	T	O	N	E	F	R	E	E
N	E	U	R		E	R	E	W	E		E	A	D	S
I	S	I				G	N	A	R		Y	M	C	A

Puzzle Solution 2

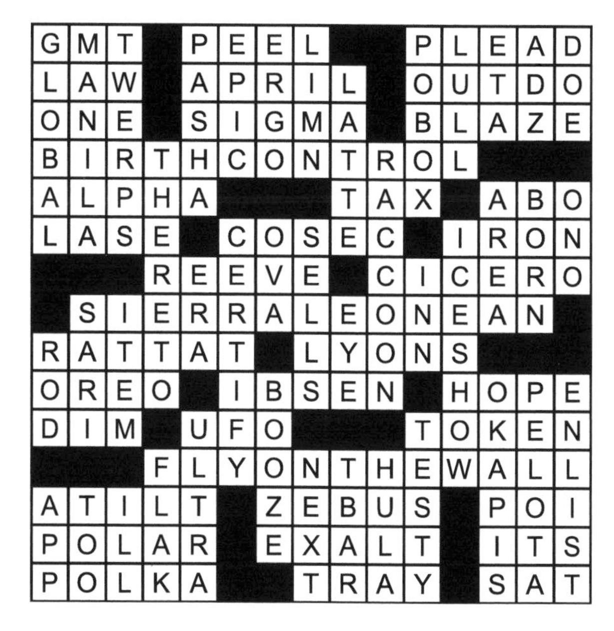

G	M	T		P	E	E	L			P	L	E	A	D
L	A	W		A	P	R	I	L		O	U	T	D	O
O	N	E		S	I	G	M	A		B	L	A	Z	E
B	I	R	T	H	C	O	N	T	R	O	L			
A	L	P	H	A				T	A	X		A	B	O
L	A	S	E		C	O	S	E	C		I	R	O	N
			R	E	E	V	E		C	I	C	E	R	O
	S	I	E	R	R	A	L	E	O	N	E	A	N	
R	A	T	T	A	T		L	Y	O	N	S			
O	R	E	O		I	B	S	E	N		H	O	P	E
D	I	M		U	F	O				T	O	K	E	N
			F	L	Y	O	N	T	H	E	W	A	L	L
A	T	I	L	T		Z	E	B	U	S		P	O	I
P	O	L	A	R		E	X	A	L	T		I	T	S
P	O	L	K	A			T	R	A	Y		S	A	T

Puzzle Solution 3

C	D	T		A	D	P			S	L	O	I	N	S
L	E	O		M	C	A	T		T	E	S	S	I	E
A	D	R		C	O	W	H		O	O	L	A	L	A
P	E	R	K	S			R	A	W	P	O	W	E	R
P	E	E	L		A	M	E	D	E	O				
		N	U	M	E	R	I	C		L	E	D	E	E
B	E	T		A	S	S	T			D	E	O	T	S
A	P	I	N	T		C	T	R		I	O	W	A	S
S	H	A	M	U			L	I	I	I		N	H	A
H	A	L	E	R		T	E	C	H	I	E	S		
				A	S	H	P	A	N		M	T	S	T
C	R	O	S	T	I	N	I			P	T	R	A	P
A	U	S	S	I	E		G	H	A	T		E	N	O
L	E	S	I	O	N		S	I	T	U		A	D	D
C	R	E	N	N	A			S	P	I		M	E	A

Puzzle Solution 4

T	A	I		B	E	A	R	D				M	B	A
E	S	T		A	L	T	E	R		S	A	F	E	R
S	H	H		T	O	O	F	A	R	A	P	A	R	T
L	E	I	B	N	I	Z		W	E	L	T			
A	N	N	E	E			J	U	D	O		S	O	O
		K	I	X	I	T	U	P	A	N	O	T	C	H
N	E	I		T	R	O	N				J	J	A	R
E	L	L	S		R	A	G	I	N		O	O	L	O
A	L	L	O				L	O	E	B		H	A	B
R	A	P	T	A	T	T	E	N	T	I	O	N		
S	S	A		D	E	W	S			P	E	S	O	S
			A	V	E	O		D	O	O	R	W	A	Y
F	E	R	R	I	S	W	H	E	E	L		O	K	S
R	A	O	U	L		A	E	T	N	A		R	I	C
O	P	S				Y	E	S	O	R		T	E	O

Puzzle Solution 5

D	Y	A	D		M	G	E	R		S	A	L	A	T
R	I	J	O		O	I	S	E		A	L	E	R	O
E	N	E	R		T	N	T	S		D	E	O	N	S
D	G	T	M	K	H	Z					X	I	I	I
				O	R	A	N	G	E	F	E	V	E	R
R	W	A	N	D	A		I	E	R	E	I			
O	A	T	S			A	C	T	I	N		T	W	O
P	U	T	A	F	I	R	E	U	N	D	E	R	I	T
O	L	S		E	S	T	O	P			C	A	S	A
			R	A	M	E	N		B	L	I	M	E	Y
T	I	M	E	R	E	L	E	A	S	E				
U	R	I	M					S	T	D	E	N	I	S
T	I	N	A	S		G	O	T	A		T	P	K	E
E	N	D	I	T		E	F	O	R		R	I	I	S
E	G	A	L	S		L	T	R	S		E	N	D	S

Puzzle Solution 6

	C	A	A		O	G	P	U		B	E	A	D	Y
K	A	T	T		R	A	I	N		A	N	T	R	A
I	N	T	E	R	B	R	E	D		D	O	T	E	S
D	A	L	A	I					N	G	W	A	Y	
D	A	E		G	E	R		S	O	U		C	F	C
O	N	E	V	O	T	E		A	N	Y		H	U	R
			A	R	C	T	U	R	U	S		E	S	E
M	S	E	C		H	Y	S	O	N		M	D	S	E
T	H	X		R	I	P	E	N	I	N	G			
H	O	T		E	N	E		G	O	I	R	I	S	H
S	E	E		R	G	S		S	N	E		N	C	O
	T	R	O	I	S					C	O	D	Y	S
O	R	N	O	N		B	I	B	L	E	B	E	L	T
R	E	A	M	S		A	R	I	B		O	B	L	A
R	E	L	E	E		T	A	G	S		E	T	A	

Puzzle Solution 7

O	L	G	A		C	O	A	L		L	O	S	T	H
H	O	L	M		O	N	C	E		I	D	E	E	S
H	O	A	R		Y	E	O	H		B	O	N	D	I
I	N	D	I	A	N	A	P	A	C	E	R			
			T	P	E	R		R	A	R		S	E	T
	L	S	A	T	S				N	T	U	P	L	E
L	A	L		S	S	O	R	C	A	Y	R	E	V	E
A	T	A	N			N	A	P			I	W	I	N
D	O	M	O	A	R	I	G	A	T	O		E	S	A
D	Y	N	A	M	O				G	L	A	D	H	
S	A	G		P	B	A		F	E	A	T			
			A	B	B	E	Y	L	I	N	C	O	L	N
L	E	O	X	I		T	A	U	S		A	D	I	O
G	E	N	O	A		A	N	T	E		M	A	R	G
S	W	A	N	N		T	A	E	L		P	S	A	S

Puzzle Solution 8

S	D	L	T	G			V	E	L	A		T	E	A
P	E	K	O	E		S	Q	U	A	B		R	R	R
H	Y	P	E	R	I	N	F	I	O	N		A	D	R
			R	I	R	E			T	E	G	N	E	R
	F	E	R		I	A	N		I	R	R	S		
O	R	L		E	N	D	O	R	A		A	P	U	
W	A	L	E	S	A		V	E	N	A	C	A	V	A
I	N	I	T	S		N	I	C		R	E	R	U	N
E	C	S	T	A	T	I	C		M	O	D	E	L	A
	O	I	O		W	H	E	L	A	N		N	A	T
		S	R	T	A		S	I	N		S	T	S	
D	E	L	E	A	D			K	O	L	A			
O	P	A		E	D	D	I	E	F	I	S	H	E	R
T	H	N		L	L	O	S	A		F	E	M	M	E
S	A	D		S	E	T	H			E	S	S	A	I

Puzzle Solution 9

M	E	D	S			B	L	A	M		U	B	R	A
A	R	A	T		B	R	O	M	O		R	E	O	S
R	I	N	G	B	E	A	R	E	R		A	R	T	A
Y	T	D		R	E	G	D		O	I	L	M	A	N
J	U	D	A	I	C			M	C	M	I			
			J	O	H	N	N	Y	C	A	C	H	E	T
L	O	C	A			A	A	R	O	N		I	O	R
A	P	E	R			T	H	R			C	S	N	E
D	E	R		U	H	H	U	H			O	N	S	E
I	C	E	C	R	E	A	M	S	O	D	A			
			S	I	A	N			O	O	L	A	L	A
M	O	T	T	S	T		N	O	N	W		C	A	R
A	W	H	O		H	U	G	H	A	N	D	C	R	Y
Y	E	A	R		E	T	O	N	S		D	E	V	A
A	D	E	E		N	E	R	O			E	L	A	N

Puzzle Solution 10

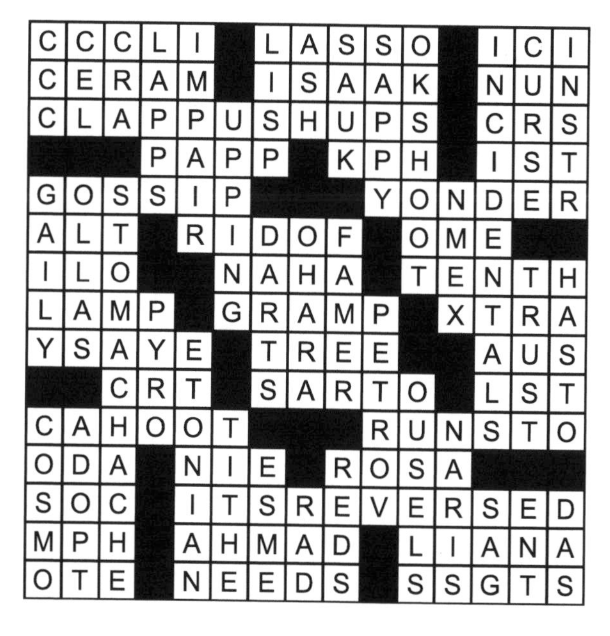

C	C	C	L	I		L	A	S	S	O		I	C	I
C	E	R	A	M		I	S	A	A	K		N	U	N
C	L	A	P	P	U	S	H	U	P	S		C	R	S
			P	A	P	P		K	P	H		I	S	T
G	O	S	S	I	P				Y	O	N	D	E	R
A	L	T		R	I	D	O	F		O	M	E		
I	L	O			N	A	H	A		T	E	N	T	H
L	A	M	P		G	R	A	M	P		X	T	R	A
Y	S	A	Y	E		T	R	E	E			A	U	S
		C	R	T		S	A	R	T	O		L	S	T
C	A	H	O	O	T				R	U	N	S	T	O
O	D	A		N	I	E		R	O	S	A			
S	O	C		I	T	S	R	E	V	E	R	S	E	D
M	P	H		A	H	M	A	D		L	I	A	N	A
O	T	E		N	E	E	D	S		S	S	G	T	S

Puzzle Solution 11

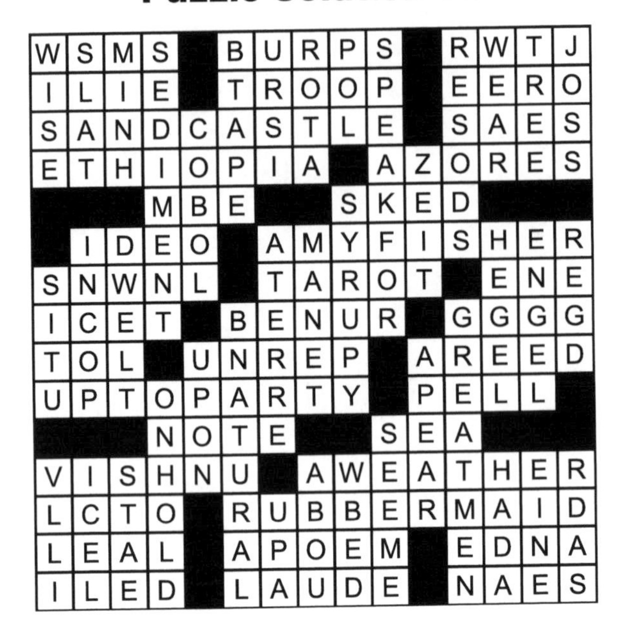

W	S	M	S		B	U	R	P	S		R	W	T	J
I	L	I	E		T	R	O	O	P		E	E	R	O
S	A	N	D	C	A	S	T	L	E		S	A	E	S
E	T	H	I	O	P	I	A		A	Z	O	R	E	S
			M	B	E			S	K	E	D			
	I	D	E	O		A	M	Y	F	I	S	H	E	R
S	N	W	N	L		T	A	R	O	T		E	N	E
I	C	E	T		B	E	N	U	R		G	G	G	G
T	O	L		U	N	R	E	P		A	R	E	E	D
U	P	T	O	P	A	R	T	Y		P	E	L	L	
			N	O	T	E			S	E	A			
V	I	S	H	N	U		A	W	E	A	T	H	E	R
L	C	T	O		R	U	B	B	E	R	M	A	I	D
L	E	A	L		A	P	O	E	M		E	D	N	A
I	L	E	D		L	A	U	D	E		N	A	E	S

Puzzle Solution 12

I	M	A	S		A	F	A	R		G	E	N	I	C
T	A	M	E		R	O	T	C		O	R	O	N	O
S	K	A	T	E	B	O	A	R	D	T	R	I	C	K
N	O	T	A	T		T	R	U	E	D		R	A	E
O	S	O		O	S	S		M	F	O	R			
		L	E	N	G	T	H	B	Y	W	I	D	T	H
			O	S	T	E	O			N	S	Y	N	C
D	C	C	L		S	P	O	S	A		E	S	N	E
K	I	W	I	S			H	I	T	A	T			
A	N	T	A	T	T	R	A	C	T	I	O	N		
			N	E	U	E		E	A	R		A	M	S
S	K	D		A	T	T	A	M		O	F	D	A	Y
H	O	T	C	R	O	S	S	B	U	N	N	I	E	S
O	B	E	L	I		Y	E	O	H		M	R	S	C
R	U	N	I	C		N	A	Y	S		A	S	T	O

Puzzle Solution 13

A	D	A	M	A		L	Y	N	X		N	A	H	A
N	O	T	O	N		A	M	I	E		O	B	E	S
I	D	T	A	G		V	I	E	D		D	O	L	L
M	O	N	T	E	C	A	R	L	O		D	O	D	I
				L	E	R		S	U	R	I	N	A	M
H	O	T	C	O	C	O	A		T	U	N			
U	R	A	L		U	C	S	B		E	G	G	O	S
A	F	E	E		M	K	U	O	G		O	O	N	A
C	E	R	A	M		S	R	T	A		F	O	U	L
			N	R	C		E	T	O	U	F	F	E	E
C	O	W	H	E	R	B		L	L	B				
A	P	S	O		I	G	N	E	S	F	A	T	U	I
T	U	T	U		T	W	O	D		T	A	R	P	T
H	M	O	S		I	G	L	U		Z	A	I	R	E
S	P	N	E		C	L	O	P		C	A	B	I	N

Puzzle Solution 14

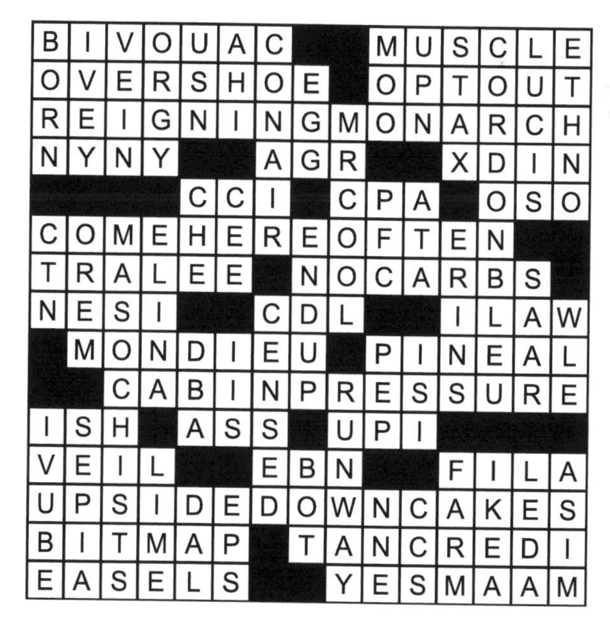

B	I	V	O	U	A	C			M	U	S	C	L	E
O	V	E	R	S	H	O	E		O	P	T	O	U	T
R	E	I	G	N	I	N	G	M	O	N	A	R	C	H
N	Y	N	Y			A	G	R			X	D	I	N
				C	C	I		C	P	A		O	S	O
C	O	M	E	H	E	R	E	O	F	T	E	N		
T	R	A	L	E	E		N	O	C	A	R	B	S	
N	E	S	I			C	D	L			I	L	A	W
	M	O	N	D	I	E	U		P	I	N	E	A	L
		C	A	B	I	N	P	R	E	S	S	U	R	E
I	S	H		A	S	S		U	P	I				
V	E	I	L			E	B	N			F	I	L	A
U	P	S	I	D	E	D	O	W	N	C	A	K	E	S
B	I	T	M	A	P		T	A	N	C	R	E	D	I
E	A	S	E	L	S			Y	E	S	M	A	A	M

Puzzle Solution 15

F	R	O	M	■	■	C	S	T	■	A	R	B	O	R
L	E	D	A	■	C	O	P	A	■	B	E	A	N	E
O	P	E	N	■	A	V	I	S	■	I	H	N	E	N
G	R	A	I	N	R	E	C	E	P	T	A	C	L	E
■	■	■	A	D	O	T	■	■	E	E	N	■	■	■
R	E	S	C	A	N	■	H	E	L	■	G	A	E	A
E	S	T	A	D	■	M	O	U	L	D	■	L	I	S
A	B	E	L	G	I	A	N	W	A	F	F	L	E	S
L	A	E	■	Y	O	R	K	E	■	I	I	I	I	I
S	T	N	S	■	W	K	S	■	E	G	G	S	O	N
■	■	■	T	B	A	■	■	B	A	H	T	■	■	■
M	O	T	I	O	N	S	T	O	S	T	R	I	K	E
L	A	H	L	X	■	S	E	C	Y	■	E	L	A	N
I	T	A	L	O	■	S	E	C	A	■	E	L	Y	S
T	H	I	E	F	■	S	D	I	■	■	S	E	S	E

Puzzle Solution 16

O	C	T	■	F	A	B	■	M	A	P	■	D	N	A
C	H	I	C	A	G	O	■	A	L	I	■	E	E	L
T	I	M	E	C	A	P	S	U	L	E	■	M	O	B
A	M	B	L	E	R	■	O	V	E	R	D	O	■	■
V	E	R	T	■	A	L	L	E	G	I	A	N	C	E
E	R	E	■	E	G	O	■	■	E	S	P	I	A	L
■	■	■	I	M	A	G	O	■	■	■	H	A	R	K
C	R	O	S	S	R	E	F	E	R	E	N	C	E	S
E	A	R	L	■	■	■	T	R	I	N	E	■	■	■
S	N	E	A	K	S	■	■	I	N	S	■	P	O	P
S	I	G	N	A	L	L	I	N	G	■	P	I	P	E
■	■	A	D	R	I	A	N	■	T	A	I	L	O	R
F	A	N	■	S	P	I	N	D	O	C	T	O	R	S
B	R	O	■	T	U	T	■	A	N	N	A	T	T	O
I	T	S	■	S	P	Y	■	B	E	E	■	S	O	N

Puzzle Solution 17

I	T	S	Y	■	A	G	H	A	■	■	C	A	B	E
N	A	H	A	■	N	L	E	R	■	U	H	L	A	N
C	H	A	L	K	D	U	S	T	■	R	I	G	I	D
R	E	Q	U	I	R	E	S	■	A	B	L	A	Z	E
■	■	■	■	D	E	O	■	E	M	A	I	L	E	D
S	Q	U	I	D	I	N	K	L	I	N	G	■	■	■
A	U	N	T	Y	■	■	U	L	L	A	■	N	O	I
M	A	D	E	■	S	C	R	E	E	■	B	O	I	L
M	H	O	■	B	A	R	T	■	■	L	I	E	S	L
■	■	■	E	L	V	I	S	P	R	E	S	L	E	Y
E	S	T	R	E	E	T	■	E	A	N	■	■	■	■
T	H	R	O	W	S	■	D	A	M	N	A	B	L	E
C	O	A	T	I	■	H	A	P	P	Y	D	A	Y	S
H	A	D	I	T	■	E	V	O	E	■	I	L	L	T
A	L	E	C	■	■	R	E	D	D	■	O	L	E	S

Puzzle Solution 18

B	N	E	R	■	U	R	G	E	■	A	V	I	E	W
A	S	E	A	■	B	I	O	L	■	N	I	N	T	H
P	E	R	L	■	U	C	S	B	■	A	L	A	T	E
T	W	O	F	O	R	T	H	E	S	E	E	S	A	W
■	■	■	■	N	O	U	■	■	O	M	S	■	■	■
C	A	L	M	D	I	S	P	O	S	I	T	I	O	N
I	M	A	G	O	■	■	N	C	A	A	■	S	F	O
T	O	I	T	■	D	A	S	D	S	■	A	T	T	N
E	S	K	■	Z	U	L	U	■	■	W	I	L	E	E
S	T	A	T	E	D	E	P	A	R	T	M	E	N	T
■	■	■	U	A	E	■	■	L	E	I	■	■	■	■
I	D	Y	L	L	S	O	F	T	H	E	K	I	N	G
G	E	S	S	O	■	C	H	A	O	■	A	A	A	A
E	M	E	A	T	■	T	U	R	N	■	E	M	P	T
T	E	R	N	S	■	A	P	S	E	■	L	I	E	R

Puzzle Solution 19

R	U	R	A	L		H	E	M	A		C	A	R	J
U	R	A	N	O		O	M	E	N		U	R	A	L
S	C	R	A	T	C	H	A	N	D	S	N	I	F	F
S	H	I	P		A	U	G	U	S	T	A			
I	I	N		A	T	M	S			B	R	D	D	A
A	N	G	E	L	A			L	A	D	D	E	R	S
			S	A	L	T	P	I	T			R	G	T
	L	A	C	E	Y	C	O	M	E	H	O	M	E	
S	E	L			S	H	I	E	L	D	S			
S	I	L	E	N	T	R			I	Q	T	E	S	T
I	S	Y	O	U			G	R	E	S		A	A	H
			N	I	P	P	I	E	R		O	R	T	O
P	O	L	I	T	I	C	A	L	S	Y	S	T	E	M
B	E	T	A		A	T	N	O		O	S	H	E	A
A	D	D	N		S	S	T	S		B	O	S	N	S

Puzzle Solution 20

H	A	S	A		N	I	C	H	T		S	P	F	S
A	B	U	G		A	S	H	U	R		L	E	I	A
G	O	B	I		N	A	O	M	I		A	T	L	I
	V	U	L	C	A	N	M	I	N	D	M	E	L	D
W	E	R	E	A			P	R	I	E		R	I	O
A	P	B		S	P	H				S	T	I	N	K
D	A	I	S		T	A	M	A	R	A	C			
	R	A	C	H	E	L	C	R	A	C	H	E	L	
			A	L	R	O	S	E	N		R	A	O	M
P	L	A	N	A				I	K	E		T	O	I
I	A	T		N	E	A	T			L	L	A	N	O
T	H	E	E	D	A	M	S	F	A	M	I	L	Y	
S	O	S	A		R	O	I	L	S		K	O	B	U
A	R	T	S		L	O	D	E	S		U	N	I	S
W	E	S	T		S	N	E	A	D		D	E	N	E

Puzzle Solution 21

J	A	M	B		A	S	T	I		W	E	S	A	Y
O	B	O	E		A	P	E	G		H	I	T	M	E
E	A	R	L		B	U	R	L		O	N	E	I	L
S	A	N	A	T	O	R	I	U	M	S		E	S	P
			B	O	N	N			O	O	M	P	H	S
M	Y	L	O	R	D		S	L	I	N	G	S		
D	O	E	R	R		B	T	E	R	F	R	I	E	D
A	U	F			A	R	A	C	E			D	D	E
S	E	T	S	A	P	A	R	T		A	P	E	E	K
		S	A	S	S	E	D		C	U	R	D	L	E
H	O	R	N	S	I			K	A	R	O			
O	N	I		I	S	T	H	A	T	A	F	A	C	T
R	I	G	G	S		H	O	P	E		E	L	M	S
S	O	H	O	T		R	Y	U	N		S	I	L	K
E	N	T	E	S		A	T	T	A		S	A	I	S

Puzzle Solution 22

M	S	E	C			R	E	A	L	M		M	T	G
F	I	N	A	L		A	E	S	I	R		O	A	R
R	E	D	H	O	T	P	O	K	E	R		S	T	E
		A	N	N	A		C	O	L	O	S	S	A	E
W	A	R		G	I	B		F	O	G	H	O	R	N
E	L	O		B	L	U	M		W	E	T			
B	T	U	S		G	R	I	A		R	E	L	I	T
M	A	N	I	F	E	S	T	D	E	S	T	I	N	Y
D	I	D	T	O		A	L	L	N		L	I	M	P
			T	L	C		A	I	T	S		P	E	E
K	N	E	E	L	E	R		B	O	R		I	M	B
M	I	R	R	O	R	E	D		M	I	S	C		
A	T	N		W	I	C	C	A	B	A	S	K	E	T
R	E	O		E	S	T	O	P		H	E	U	R	E
T	R	S		R	E	I	N	S			E	P	I	C

Puzzle Solution 23

H	O	N	E		J	E	U	L	R		N	E	G	S
E	T	O	N		A	S	P	I	E		O	N	U	E
P	O	W	D	E	R	P	U	F	F		B	O	I	L
C	H	A	S	M			P	E	L	L	I	C	L	E
			U	B	R	A			O	U	G	H	T	S
F	L	I	P	O	N	E	S	T	W	I	G			
L	E	S		W	A	N	N	A		S	I	G	N	S
U	N	I	E		S	E	W	U	P		E	E	O	C
B	A	N	D	B		I	N	T	O	W		R	N	A
			M	I	D	D	L	E	L	E	T	T	E	R
N	E	R	O	L	I			R	O	W	E			
C	A	R	N	E	G	I	E			O	H	G	E	E
A	S	S	T		G	O	L	D	E	N	R	U	L	E
A	T	T	O		E	W	O	K	S		A	N	E	G
S	M	A	N		R	A	N	A	T		N	G	A	S

Puzzle Solution 24

M	A	D	A	S		L	A	N	I		I	M	H	O
O	Z	A	W	A		E	S	A	S		F	E	A	R
L	A	U	E	R		P	T	U	I		C	T	R	L
E	L	D	D	A	S	E	H	T	N	I		H	A	E
S	E	E			O	W	E	S		S	H	A	K	
T	A	T	A	M	I				S	H	A	N	I	A
			H	O	G	W	A	S	H		L	O	R	E
P	R	I	C	I	N	G	C	A	T	H	O	L	I	C
E	A	R	H		E	T	A	G	E	R	E			
G	R	O	O	V	E				T	E	S	T	E	D
	A	N	O	A		D	A	N	L			B	L	O
S	A	W		T	H	I	N	G	S	W	R	O	N	G
A	V	O	N		R	C	A	F		A	R	N	I	E
B	I	O	G		A	T	M	O		B	U	E	N	A
U	S	D	O		S	U	E	R		E	N	S	O	R

Puzzle Solution 25

W	A	D	I		E	F	R	E	M		S	L	A	G
O	R	A	D		G	I	O	C	O		T	E	R	M
L	E	W	I	S	O	F	P	O	P	M	U	S	I	C
F	A	N	G	B	I	T	E			E	C	H	E	S
				A	S	H		U	S	O	C			
E	X	P	E	C	T	S	T	H	E	W	O	R	S	T
B	A	R	A	K			E	R	N	S		A	C	A
E	X	I	T		H	O	L	Y	G		H	K	I	R
R	E	M		J	I	V	E			F	I	E	F	S
T	S	A	R	A	L	E	X	A	N	D	E	R	I	I
			A	M	O	R		M	O	T				
O	M	A	N	I			H	O	T	D	A	T	E	S
T	O	A	L	E	S	S	E	R	E	X	T	E	N	T
B	A	R	O		C	E	N	T	R		U	R	G	E
S	T	E	W		S	S	S	S	S		B	A	S	T

Puzzle Solution 26

A	S	T	A	B		E	P	P	S		S	P	F	S
B	A	H	I	A		M	I	L	T		Q	U	I	K
O	D	E	R	N	E	I	S	S	E		U	N	D	E
U	A	R		K	E	G			R	R	A	T	E	D
		M	A	J	O	R	G	E	N	E	R	A	L	S
T	W	O	T	O		E	N	D	U	S	E			
F	A	S	T	B			O	N	M	E		A	V	I
R	Y	E	S		U	N	C	A	S		H	B	A	G
S	S	T		U	S	O	C			B	U	S	C	H
			O	V	E	R	H	S		I	T	S	A	T
M	I	S	T	E	R	N	I	C	E	G	U	Y		
A	N	I	T	A	S			A	D	C		S	C	S
R	A	T	A		F	A	N	G	T	A	S	T	I	C
I	L	A	W		E	M	A	G		T	H	E	T	A
A	L	T	A		E	B	B	S		S	U	M	O	S

Puzzle Solution 27

A	I	W	A		A	Z	E	R	I		I	M	E	T
M	G	R	S		R	I	C	A	N		M	A	M	A
P	O	W	D	E	R	P	U	F	F		P	N	E	U
E	T	C		T	E	A	S	E	L		A	O	U	T
D	A	K	O	T	A				U	C	L	A		
			P	A	R	L	O	R	X	G	A	M	E	S
O	G	E	E	S		I	S	S	E	I		A	X	L
V	E	N	N		O	L	A	V	S		E	N	E	O
E	N	T		A	L	A	M	P		R	E	O	R	G
N	E	A	R	D	I	S	A	S	T	E	R			
		I	R	O	V				A	P	O	L	A	R
G	A	L	A		E	L	A	P	S	E		A	N	E
A	L	I	T		O	L	L	I	E	N	O	R	T	H
R	E	N	E		I	D	L	E	R		N	C	A	A
R	E	G	D		L	S	A	T	S		S	H	E	B

Puzzle Solution 28

S	U	R	G		E	N	D	S		P	A	W	A	T
A	S	I	A		R	O	S	A		O	V	E	R	T
V	I	N	G	R	O	O	M	S		E	E	R	I	E
E	A	G	L	E	T			E	I	T	R	E	A	N
			A	N	I		S	S	S	S	S			
B	R	O	W	N	C	O	W		T	C	E	L	L	S
A	O	N		E	A	T	E	R	S		T	I	E	A
S	M	E	L	T		H	N	C		A	O	K	A	Y
T	A	T	I		M	O	S	A	I	C		E	S	T
A	N	O	M	I	E		O	S	T	I	N	A	T	O
			A	T	O	W	N		S	D	I			
C	A	M	P	A	R	I			E	I	T	H	E	R
L	A	N	E	S		L	A	R	A	C	R	O	F	T
A	B	O	R	C		C	A	S	S		I	O	L	E
P	A	P	U	A		O	A	T	Y		C	P	A	S

Puzzle Solution 29

A	B	E	L	S		P	A	L	P	S		P	T	A
G	E	N	I	C		A	R	O	S	E		U	H	F
H	E	A	V	Y	S	L	I	G	H	T		R	A	X
A	R	T	E	L	S		B	I	A	S	T	I	R	E
S	S	E		L	H	Z			W	O	R	T		
			A	A	A	U	G	H		N	Y	A	L	A
H	T	M	L		P	L	I	E	S		S	N	I	P
O	H	O	S		E	E	N	I	E		A	I	M	S
L	A	L	O		D	M	A	S	S		I	C	E	E
T	W	E	R	P		A	S	T	A	L	L			
		H	A	E	C			S	M	A		S	T	O
C	O	I	N	S	L	O	T		E	T	A	L	I	A
H	U	L		E	X	H	A	U	S	T	F	A	N	S
O	I	L		T	V	S	E	T		E	E	M	C	T
U	S	S		A	I	O	L	I		N	E	S	T	S

Puzzle Solution 30

I	N	S	Y	N	C		K	I	D	S		S	T	A
S	O	S	O	O	N		A	R	I	P		P	O	S
L	O	S	T	I	N	S	P	A	C	E		A	S	L
A	S	S	E	T		S	P	I	K	E		C	C	I
M	E	S	A		S	E	A	L	E	D	B	E	A	M
			M	G	T				Y	O	U	R	N	
S	P	O	O	N	R	E	S	T			M	A	I	S
T	O	L		L	A	B	E	L	L	E		C	N	S
U	P	D	O			B	E	C	A	U	S	E	I	T
	E	F	L	A	T				D	D	E			
A	L	L	E	V	I	A	T	E	D		T	O	R	A
L	E	A		E	R	L	E	S		E	S	S	A	I
L	O	M		R	A	I	S	E	S	A	H	A	N	D
A	X	E		T	N	T	S		S	T	O	G	I	E
H	I	S		S	A	Y	A		R	A	T	E	D	R

Puzzle Solution 31

Puzzle Solution 32

L	S	T		A	C	E	H		C	R	A	D	L	E
Y	O	K	O	H	A	M	A		R	E	P	O	U	R
S	P	L	E	N	D	I	D		I	S	E	U	L	T
O	P	E	D		G	R	A	B		I	M	B	U	E
L	Y	S	I	N	E		C	U	D	G	E	L		
			P	I	S	H		M	I	N	N	E	H	A
S	E	D	U	M		U	P	S	Y			H	A	S
T	R	E	S	H		M	A	O		H	T	E	S	T
E	N	A			B	A	K	U		A	O	R	T	A
N	I	L	S	S	O	N		T	V	M	A			
		T	O	T	A	L	S		I	M	H	E	R	S
A	B	O	M	A		Y	U	C	K		A	D	A	H
S	A	V	A	N	T		S	A	I	L	I	N	T	O
Y	I	E	L	D	S		I	N	N	E	R	A	C	E
E	N	R	I	C	O		E	E	G	S		S	H	S

Puzzle Solution 33

I	N	S	P		S	T	R	U	M		T	A	E	L
L	A	I	R		T	H	I	S	I		E	I	N	E
I	N	C	O	R	R	E	C	T	C	H	A	N	G	E
A	G	A	P	E		A	S	O		O	A	T	E	R
			H	I	S	S			R	A	C	I	L	Y
A	B	J	E	C	T	P	O	V	E	R	T			
S	R	I	S	E	S		B	U	D	D		V	E	R
E	E	L	Y		I	D	E	E	S		R	E	L	O
C	E	L		A	M	O	Y		T	E	E	N	I	E
			K	N	O	T	S	L	A	N	D	I	N	G
S	U	S	A	N	N			E	R	D	E			
A	N	E	Y	E		S	S	S		E	M	M	A	S
R	E	P	O	S	S	E	S	S	E	D	A	U	T	O
A	R	T	E		S	C	I	O	N		N	S	E	W
S	P	A	D		N	O	N	N	A		D	E	M	S

Puzzle Solution 34

A	B	B	E		A	K	I	M		I	L	L	E	R
B	E	R	N		L	O	M	A		W	A	I	V	E
D	E	A	D		B	O	A	T	R	A	I	L	E	R
U	R	S	U	L	A			C	O	S		T	R	E
L	O	S	E	O	N	E	S	H	E	A	D			
	N	T	S	B		M	O	B			I	M	A	M
A	T	A		A	S	I	P		M	C	C	A	B	E
B	A	C	K	T	O	T	H	E	F	U	T	U	R	E
U	P	K	E	E	P		I	T	A	R		R	O	K
Y	E	S	T			P	S	Y		S	A	I	N	
			T	R	U	S	T	M	E	O	N	T	X	Y
G	I	A		I	B	A			O	R	N	A	T	E
I	R	R	E	G	U	L	A	R	S		A	N	A	T
F	O	R	E	H		M	M	C	I		L	I	L	I
S	C	A	N	T		E	B	A	N		S	A	E	S

Puzzle Solution 35

S	P	I	N	■	A	N	O	D	E	■	I	T	N	O
C	L	I	O	■	S	O	L	I	D	■	T	R	E	N
H	A	I	R	Y	C	H	E	S	T	■	S	A	B	O
M	T	I	D	A	■	I	O	S	■	I	A	G	O	S
O	H	I	■	W	E	T	S	U	I	T	S	■	■	■
■	■	■	A	L	M	S	■	A	N	G	E	L	O	S
T	A	I	L	S	P	■	A	D	V	O	C	A	T	E
I	S	A	T	■	T	E	P	E	E	■	R	O	O	M
N	O	S	E	D	I	V	E	■	R	E	E	S	E	S
S	P	I	R	A	E	A	■	N	T	W	T	■	■	■
■	■	■	N	A	R	C	I	S	S	I	■	U	P	A
M	C	R	A	E	■	U	N	I	■	N	A	N	A	S
A	L	I	I	■	B	A	G	G	A	G	E	C	A	R
S	E	R	V	■	E	T	O	N	S	■	N	A	V	E
S	M	E	E	■	G	E	T	S	L	■	A	P	O	D

Puzzle Solution 36

B	A	S	S	■	Q	U	A	K	E	■	S	I	D	E
R	I	T	E	■	U	L	C	E	R	■	E	N	I	D
O	N	E	A	N	O	T	H	E	R	■	A	C	E	D
S	U	M	M	I	T	■	E	N	O	R	M	I	T	Y
■	■	■	A	S	A	P	■	■	R	U	E	D	■	■
I	N	A	N	E	■	O	P	T	■	D	R	E	S	S
B	I	B	L	I	O	P	H	I	L	E	■	N	T	H
E	E	R	Y	■	O	L	O	G	Y	■	S	T	O	A
A	C	E	■	T	H	I	T	H	E	R	W	A	R	D
M	E	A	N	S	■	N	O	T	■	H	A	L	E	Y
■	■	C	I	A	O	■	■	S	P	E	D	■	■	■
A	S	T	E	R	I	S	K	■	L	A	D	D	I	E
C	H	I	C	■	L	O	N	G	I	S	L	A	N	D
M	A	N	E	■	E	R	O	S	E	■	E	D	D	A
E	G	G	S	■	D	E	B	A	R	■	D	O	O	M

Puzzle Solution 37

E	L	S		A	E	C	I	A		S	T	R	E	P
A	A	M		S	P	I	T	Z		P	O	E	S	Y
S	U	A		T	O	A	S	T	M	A	S	T	E	R
T	R	L		A	N	O	A		E	R	E	S	S	O
M	A	L	O	R	Y	S		C	R	E	E			
		H	T	T	M		C	H	E	M		B	S	S
E	N	O	T	E		P	O	L	L	E	N	A	T	E
B	L	U	E		O	H	B	O	Y		A	L	O	P
B	A	R	R	Y	M	O	R	E		U	R	A	W	A
S	T	S		E	A	T	A		E	N	C	L		
			A	W	H	O		S	E	T	S	A	I	L
C	U	E	S	T	A		D	A	Y	O		I	N	I
I	N	C	A	R	N	A	T	I	O	N		K	B	S
N	I	C	H	E		D	E	E	R	E		A	U	T
C	E	L	I	E		O	N	R	E	D		S	D	S

Puzzle Solution 38

O	N	I	C	E		B	A	D	S		B	I	B	I
S	E	N	A	T		E	R	I	U		O	N	I	T
M	A	L	T	E	S	E	C	A	T		M	T	G	S
I	R	A	S		H	C	H		R	I	B	E	Y	E
U	T	I		N	T	H		L	A	C	E	R	T	A
M	O	D	U	L	E		C	A	S	S		C	A	S
			C	A	T	N	A	P			R	O	X	Y
	E	A	S	T	L	A	N	S	I	N	G	M	I	
A	V	I	D			B	Y	E	N	O	W			
L	E	R		I	T	O	O		S	T	J	O	H	N
A	R	B	O	R	I	O		H	O	S		C	E	O
S	M	A	R	T	S		N	A	L		H	E	R	D
K	O	L	N		A	B	O	V	E	B	O	A	R	D
A	R	L	O		N	U	D	E		R	O	N	E	E
N	E	S	T		E	M	I	R		A	F	O	N	D

Puzzle Solution 39

M	T	S	I	N	A	I	■	T	B	S	■	S	S	A
P	A	T	M	O	R	E	■	A	A	S	■	P	P	D
S	T	U	F	F	E	D	S	O	L	E	■	O	I	O
■	■	D	I	E	M	■	O	I	L	■	L	I	G	N
S	Y	D	N	E	Y	A	U	S	T	R	A	L	I	A
R	T	E	E	■	■	T	T	T	■	E	V	E	R	I
A	D	D	■	I	E	O	H	■	B	G	I	R	L	S
■	■	■	T	H	U	M	B	H	O	L	E	■	■	■
R	A	S	H	A	D	■	E	U	D	E	■	G	U	T
I	L	I	A	D	■	O	N	T	■	■	Z	U	L	U
P	O	S	T	A	G	E	D	U	E	S	T	A	M	P
T	H	E	S	■	A	N	I	■	D	O	I	N	■	■
I	A	N	■	A	M	O	N	T	I	L	L	A	D	O
D	O	O	■	L	M	N	■	K	N	E	E	C	A	P
E	E	R	■	S	A	E	■	L	A	S	S	O	E	S

Puzzle Solution 40

F	A	S	O	■	C	A	B	O	T	■	T	P	E	R
A	L	O	W	■	O	D	E	T	S	■	S	A	L	A
S	A	I	N	T	M	I	S	B	E	H	A	V	I	N
T	I	E	G	A	M	E	S	■	■	T	R	E	N	D
■	■	■	O	N	I	T	■	O	N	E	I	D	A	S
N	O	B	A	K	E	■	V	E	I	N	S	■	■	■
P	H	I	L	E	■	A	S	I	P	■	T	F	A	L
I	T	D	■	D	E	L	I	L	A	H	■	I	L	E
N	O	E	S	■	L	A	G	S	■	E	G	A	L	E
■	■	■	H	A	Y	D	N	■	T	N	O	T	E	S
G	A	T	E	W	A	Y	■	G	I	R	O	■	■	■
A	W	A	K	E	■	■	K	A	T	Y	D	I	D	S
L	I	V	E	D	I	N	I	L	L	I	N	O	I	S
A	R	I	L	■	P	E	L	E	E	■	I	T	E	N
S	E	S	S	■	S	A	L	A	D	■	K	A	T	S

Puzzle Solution 41

	S	N	A		D	A	T	A		D	A	S	D	S
B	I	O	N		C	A	Y	S		A	P	P	I	A
P	R	I	S	O	N	E	R	S		T	E	A	S	E
E	R	D	A	S					S	S	G	T	S	
E	E	E		S	I	S		T	C	U		T	O	A
N	E	A	T	E	S	T		E	O	N		E	L	S
			E	T	I	C	K	E	T	S		R	V	S
S	O	M	A		T	R	E	T	S		A	S	E	T
A	N	A		A	S	O	N	E	M	A	N			
M	S	S		L	A	I		R	A	G	O	U	T	S
I	C	S		T	F	X		S	N	I		L	H	Z
	R	E	N	E	E					T	A	T	E	S
J	E	U	L	R		V	E	R	S	A	T	I	L	E
R	E	S	E	E		I	D	D	O		E	M	I	L
S	N	E	R	D		M	A	Y	S		S	O	P	

Puzzle Solution 42

T	Z	Z	T		I	C	E	L		S	A	T	E	D
W	A	B	E		N	O	G	O		O	T	H	E	R
I	S	A	N		S	O	O	T		U	T	I	L	E
T	U	R	N	T	U	R	N	T	U	R	N			
			I	R	R	S		A	S	I		P	C	B
	E	S	S	I	E				O	S	T	E	A	L
D	N	H		G	R	E	E	N	C	H	E	E	S	E
E	F	A	X			T	E	E			G	P	A	S
B	A	K	E	R	S	D	O	Z	E	N		E	V	S
T	I	E	R	R	A				L	E	A	D	A	
S	T	R		A	W	E		S	I	R	S			
			I	T	S	O	U	T	T	A	H	E	R	E
V	A	L	L	I		L	O	R	I		O	S	I	P
A	L	I	E	N		I	M	U	S		R	C	M	P
R	E	T	A	G		C	O	M	M		T	E	E	S

Puzzle Solution 43

C	R	O	C	K			L	A	D	A		L	M	N
S	A	C	R	E		E	D	I	E	S		E	E	E
A	S	T	O	N	M	A	R	T	I	N		T	N	T
			F	O	O	T			T	E	N	S	E	S
	M	S	T		V	I	Z		I	R	I	D		
S	E	M		S	E	N	I	L	E		C	O	M	
T	R	A	D	E	D		G	O	S	P	E	L	E	R
A	L	L	E	R		T	Z	U		I	S	U	L	I
G	O	L	G	O	T	H	A		B	U	T	N	E	T
	T	E	R		E	I	G	H	T	S		C	E	E
		R	E	I	N		S	U	E		A	H	S	
R	E	H	E	M	S			G	A	R	R			
I	N	A		P	E	T	S	E	M	A	T	A	R	Y
C	C	L		E	S	T	A	R		T	E	N	T	O
A	L	F		I	T	T	O			S	L	U	E	D

Puzzle Solution 44

T	N	U	T			S	U	C	H		C	O	N	C
I	P	S	A		A	L	C	E	E		A	L	E	S
S	E	A	N	O	C	A	S	E	Y		S	E	A	U
C	P	I		W	O	M	B		D	S	H	A	R	P
H	A	N	S	O	M			B	A	T	E			
			D	W	A	Y	N	E	Y	A	W	N	E	D
D	O	P	A			E	L	I	S	E		O	L	E
O	N	E	K			S	A	R			I	L	Y	A
H	U	R		H	A	I	K	U			S	A	S	E
A	P	T	I	T	U	D	E	T	E	S	T			
			M	E	G	O			R	O	O	F	E	R
V	I	S	H	N	U		S	A	I	L		E	L	E
I	D	E	O		S	A	N	D	C	A	S	T	L	E
L	O	A	M		T	H	I	G	H		L	A	I	C
E	L	S	E		A	L	P	S			A	S	S	E

Puzzle Solution 45

T	U	B	A	L		E	S	B	A	T		R	C	S
O	N	I	C	E		E	Q	U	U	S		E	O	E
T	I	C	O	N	D	E	R	O	G	A		A	R	G
			I	N	S	S		N	E	R		L	F	O
U	N	K	N	O	T				R	I	C	T	U	S
N	O	I		N	H	L	E	R		N	C	R		
A	T	E			E	U	D	E		A	C	O	O	L
P	A	V	E		I	N	I	T	S		I	O	D	E
T	S	I	D	E		A	L	O	P			P	E	G
		N	U	L		R	E	P	E	G		E	L	G
B	O	C	C	I	E				N	E	A	R	L	Y
R	A	L		T	R	A		S	C	O	T			
I	T	I		I	N	C	O	H	E	R	E	N	C	E
C	E	N		S	O	A	V	E		G	N	A	R	L
E	S	E		M	S	D	O	S		E	T	A	I	L

Puzzle Solution 46

R	A	B	B	I		O	O	O	H		S	C	A	B
A	C	E	I	N		U	P	S	Y		C	A	T	O
N	O	A	D	S		T	E	M	P		O	O	R	T
K	I	N	D	O	F	B	L	U	E		T	R	A	S
	N	S	Y	N	C			N	S	E	C			
				G	A	M	E	D		F	H	O	L	E
S	E	R	V		R	A	P		I	T	E	M	E	D
O	D	I	E		D	M	A	S	S		G	A	N	G
L	I	B	R	A	S		C	P	S		G	R	A	Y
I	N	S	T	R		P	T	R	A	P				
			E	D	I	E			I	L	I	A	C	
G	A	R	B		N	E	W	E	D	I	T	I	O	N
P	U	R	R		A	K	E	Y		N	I	M	O	Y
A	N	N	A		N	E	A	R		T	S	E	T	S
S	T	A	E		E	R	N	E		H	I	R	E	E

Puzzle Solution 47

G	A	N	G		C	O	A	T	I		B	T	W	O
I	R	A	E		M	I	M	I	C		A	H	A	B
T	U	R	N	S	L	O	O	S	E		D	R	G	E
A	T	Y	O	U			C	H	E	E	S	I	E	R
			A	N	D	S	O		S	P	A			
E	D	D		B	Y	E				O	N	T	A	P
L	E	I			E	S	K		O	C	T	U	L	E
I	N	C	O	R	R	E	C	T	C	H	A	N	G	E
S	T	E	P	E	S		S	U	A			I	A	T
T	E	D	I	U				R	L	S		C	E	E
			N	E	W		O	F	A	G	E			
A	G	U	I	L	E	R	A			N	T	E	S	T
V	E	L	O		W	O	R	S	T	Q	U	A	I	S
E	D	E	N		O	D	E	A	R		D	R	N	O
O	S	E	S		N	E	D	D	A		E	P	O	S

Puzzle Solution 48

R	E	D	Q		B	G	D	V			A	T	T	S
O	U	L	U		A	L	O	O	N		V	S	O	P
E	L	I	E		N	U	T	R	I		O	O	N	A
G	A	L	L	W	I	T	H	T	H	E	W	I	N	D
			L	A	S	E				I	S	L	E	S
M	M	D		C	H	I	N	O	O	K				
Y	O	U	O	K			E	C	I		R	E	A	M
D	I	R	T	Y	R	O	T	T	E	N	E	G	G	O
C	L	U	B		A	F	T			E	A	G	R	E
				G	R	A	Y	I	S	H		Y	A	T
D	I	A	N	A				T	I	R	O			
D	O	N	T	M	O	V	E	A	M	U	S	C	L	E
D	U	D	E		C	A	R	I	B		T	A	Y	E
D	S	O	S		H	Y	E	N	A		E	A	R	N
D	A	R	T			A	I	T	S		O	N	E	S

Puzzle Solution 49

N	A	S	A	L		A	H	I	G	H		D	F	C
T	A	L	I	A		T	I	T	L	E		A	A	U
H	A	R	R	Y	P	O	T	T	E	R		E	R	R
			B	I	E	R	S		N	E	E	D	B	E
M	T	D	A	N	A			A	N	I	S	E	E	D
L	E	N	G	T	H		A	S	M	A	S	H		
L	E	A		O	E	I	L	S		M	A	T	C	H
E	U	R	E		N	S	T	A	R		I	M	R	E
S	P	E	N	D		L	A	I	U	S		O	E	R
		S	O	O	N	E	R		S	E	C	R	E	T
S	E	E	T	O	I	T			T	L	O	F	P	Z
G	R	A	E	M	E		T	O	I	L	E			
A	R	R		I	N	N	E	R	C	I	R	C	L	E
T	E	C		N	T	E	S	T		N	C	A	A	S
E	D	H		G	E	N	T	S		G	E	A	R	S

Puzzle Solution 50

F	R	E	E	H		C	H	I	P		A	V	E	R
T	E	S	S	A		H	A	L	O		T	I	S	E
P	L	T	T	K		R	A	I	L		R	E	C	T
			H	I	Y	O	G	O	L	D	A	W	A	Y
S	O	W		M	O	M			S	H	L	E	P	P
P	R	O	A		D	A	M	A	T	O		D	E	E
N	O	R	I	A			E	B	E	R	T			
		D	R	A	W	S	T	A	R	S	O	N		
			H	A	I	K	U			E	M	A	L	L
A	P	B		U	N	I	P	O	D		B	R	E	W
T	H	E	A	G	E			R	W	Y		A	S	P
T	O	U	G	H	C	O	O	K	I	E	S			
E	B	L	A		A	T	A	N		G	R	O	U	T
S	I	A	M		S	E	S	E		G	I	M	L	I
T	A	H	E		K	A	T	Y		S	S	G	T	S

Printed in Great Britain
by Amazon